THE COSMOS EXPLAINED

THE COSMOS EXPLAINED

A HISTORY OF THE UNIVERSE FROM ITS BEGINNING TO TODAY AND BEYOND

CHARLES LIU

ILLUSTRATIONS BY
MAKSIM MALOWICHKO

IVY PRESS

Inspiring | Educating | Creating | Entertaining

Brimming with creative inspiration, how-to
projects, and useful information to enrich your
everyday life, quarto.com is a favorite destination
for those pursuing their interests and passions.

First published in 2022 by Ivy Press,
an imprint of The Quarto Group.
The Old Brewery, 6 Blundell Street
London, N7 9BH,
United Kingdom
T (0)20 7700 6700
www.quarto.com

Design © 2022 Quarto
Text © 2022 Charles Liu

A catalogue record for this book is available
from the British Library.

ISBN 978-0-7112-5274-5
Ebook ISBN 978-0-7112-5275-2

10 9 8 7 6 5 4 3 2 1

Design by Kevin Knight

Cover illustration by Ted Jennings

Inside illustrations by Maksim Malowichko,
except where otherwise stated on page 192.

Printed in China

MIX
Paper from
responsible sources
FSC® C016973

Contents

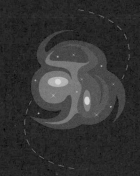

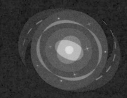

Introduction

The beginning. The past. The present. The future. The end. We humans are creatures of time, bound by the limits of before and after. We always want to hear entire stories from start to finish. We humans are also creatures of curiosity, endlessly asking questions about every part of every story. We search continually for answers, and pause only when we obtain a satisfactory explanation – for a moment, anyway, until inevitably we inundate ourselves with a fresh flood of questions.

These two great impulses – to want to know the whole story, and to need to know the reasons behind it – come crashing together in the telling of the ultimate tale: the story of everything that exists. The universe contains all of space, matter, energy and time! In our limited lives, it may seem optimistic to think we can learn anything more than a tiny bit about the cosmos, never mind that we might contemplate the whole thing all at once. We exist in the present; can we possibly hope to see through the past towards the cosmic origin, or into the future towards the cosmic finale?

Amazingly, we can and we have. Just as we humans have written down our history for future generations to read, the universe has left records, too. All we have to do is to learn the languages

in which they were written. The scripts of science – the prose of mathematics and the poetry of astronomy, physics, chemistry and biology – are how we can explain the cosmos.

In the theory of relativity, Albert Einstein explained that every interval of time in the universe has a start and an end. Cosmic history is thus a vast web of beginnings and endings. Our own strand of it runs unbroken from the Big Bang and the expansion of space to the separation of forces, the creation of matter, the formation of stars and galaxies, and then the birth of the Milky Way, the Sun, Earth, life and us. With science we can look

far ahead in time as well, and predict an end to the stars and planets and perhaps ultimately even black holes and all of matter itself. The universe will then live on, dark and still, until – what? Although for now, we only see the future through a glass darkly, the forward progress of science will bring us ever clearer vision. Hopefully, then, someday we shall know.

And the present? It is today. You are here. I am honoured and humbled to share with you the story of the cosmos, explained thanks to the combined work of scientists throughout human history, and told here from its beginning to this moment in time – and beyond.

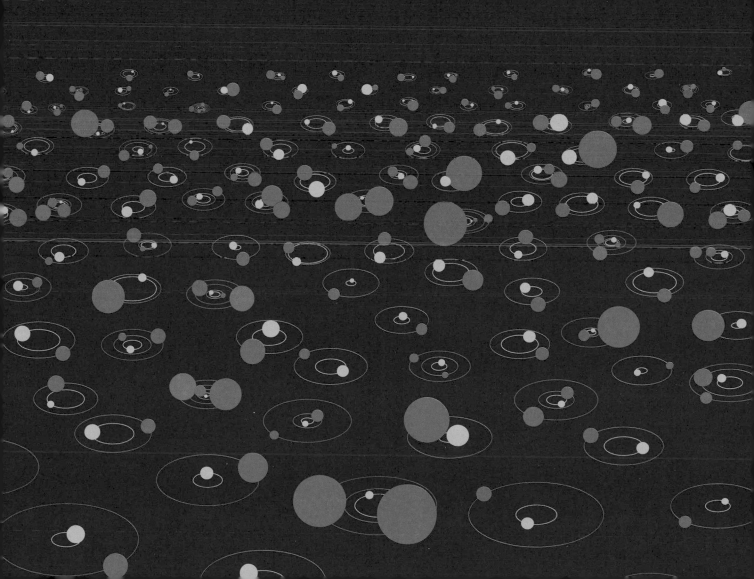

0 seconds

1

The First Quadrillionth of a Quadrillionth of a Second

The Cosmos is Born

A birth, as a mother will tell you, is momentous, miraculous – and messy.

The first instants of a life are filled with activity that should follow a reasonably predictable order of events, but always winds up wrapped in the utter chaos of unexpected occurrences that seem to arise from nowhere. In the blink of an eye, something that does not have its own existence comes into being – and history will never be the same again.

The birth of time has exactly this character, although it's wilder by far than our own modest beginnings. At the origin of everything, the life of the universe begins with a dazzling, explosive expansion – the Big Bang. The laws of physics predict that the expansion should continue without interruption, at a smoothly changing rate; but then, an unpredicted infusion of energy, seemingly coming from nowhere, inflates the universe almost beyond recognition.

The nascent patterns in the dense energy within space and time freeze into place. The universe eventually settles back into smooth expansion – but not before its size, shape, character and history have been irreversibly transformed as the cosmos comes into existence.

One huge difference, however, contrasts the beginning of a person's life and the beginning of the cosmos. Time, space, matter, energy and the rules of nature that govern them exist while a human baby is a foetus, or an embryo, or even a blastocyst. There is, on the other hand, no prenatal phase for the universe – there is no 'before'. So what made time begin to tick? Nothing comes from nothing. Or does it? Could the universe truly have arisen from a void?

Incredibly, we must look beyond time and space for the answer. It boggles the mind to have to get to grips with the idea that the universe arose from an environment where our current laws of physics do not apply – yet it is the best way, and maybe the only way, to make sense of it all. With our limited grasp of what is and can be, we return to our own human birth process for analogies. A pre-person is nurtured in a womb; was there a super-dimensional multiverse that provided the conditions to produce our universe? An event – perhaps a set of events and conditions – causes one special human cell to grow, multiply and ultimately develop into a unique human being. What event or conditions could possibly make a universe?

How fitting that we should begin the timeline of the universe with some of the most challenging mysteries that humanity has ever faced. Undaunted, we do our best to apply the tools and methods of science – the remarkably successful system we have developed over the centuries to seek answers about the natural world – to the very start of cosmic time. So far, we have many more questions than we have answers – exactly what we'd expect at the earliest possible edge of discovery.

The Planck Time

At the moment of the Big Bang, the universe had zero volume and infinite density.

Such an object is called a singularity, and the laws of nature as we understand them don't work in it. How long was it, then, before the laws of nature did start to work?

We don't yet know for sure, but we know of an upper limit to that time interval. We are guided by two fundamental scientific theories: quantum theory and the general theory of relativity. Each theory has a length scale, called the Compton wavelength and the Schwarzschild radius respectively, within which certain fundamental measurements are impossible. When those limits match one other in a single object, they are each about 10^{-35} metres – 1 hundred-millionth of a trillionth of a quadrillionth of a metre – in size and the time that it would take a beam of light to span that distance is just 10^{-43} seconds – 1 ten-millionth of a billionth of a trillionth of a quadrillionth of a second!

That almost unimaginably short duration is called the Planck time, after the German physicist Max Planck. Whatever happened during that earliest moment in time can't be explained by any of physical rules we currently know. Nevertheless, it set up everything that has happened since then in the history of the cosmos.

The general theory of relativity describes the motion of objects through space and time in the universe. First published by Albert Einstein in 1915, the theory shows how space isn't just empty nothingness, but rather a flexible medium like a jelly that can bend, dimple or even twist. Things with mass curve space towards them; that curvature is gravity, which changes the motion of objects just as if they're being pulled by a force. More massive objects (such as stars or planets) curve space more than less massive objects (such as people or pebbles) and thus create more gravity. Perhaps most amazingly, space and time are tied together into a four-dimensional fabric called space-time; thus, time is a dimension, much like length, width and height, just with different properties.

Quantum theory describes the behaviour and interaction of the tiny particles that comprise all of the matter and energy in the universe. Developed during the first decades of the twentieth century, the theory shows how atoms and molecules absorb and release energy only in specified amounts known as quanta, with different substances exhibiting different patterns of quanta depending on their properties and their surroundings. Quantum theory also describes how energy and matter can be both particles and waves – so, for example, a beam of light can be both a stream of energetic particles and an energy-carrying wave at the same time.

BIOGRAPHY

Max Planck (1858–1947) was a German physicist who, in 1900, discovered a solution to a problem about radiation from warm objects that used the concept of light quanta. His work pioneered the field of quantum mechanics. Today, Germany's leading institution of scientific research is named the Max Planck Society in his honour.

0.0000000000000000000000000000000000000000001 (10⁻⁴³) seconds

The Big Bang – What Happened

The Planck time has elapsed. The Big Bang begins in earnest now.

The Big Bang isn't an explosion that occurs in the universe; it is the explosion of the universe. Whatever happened within that tiny window of time between 0 and 10^{-43} seconds to cause the liberation of space has done its work – and now the volume of the cosmos is growing, starting at the Planck scale of 10^{-35} metres and getting larger with each passing instant. Every bit of space that exists today was once encompassed in that tiny kernel; so today, if someone asks you where the Big Bang happened, you can say: 'Here, there and everywhere'.

The laws of physics as we experience them today start to apply to the cosmos now. One property that begins to matter immediately is temperature – a measure of the average amount of randomly directed motion and vibration contained in a system. Astronomers prefer to use the Kelvin temperature scale, named for Lord Kelvin (or more precisely, William Thomson, Baron Kelvin, 1824–1907), on which zero is the lowest possible temperature in the universe and a change of 1 Kelvin is equivalent to a change of 1 degree Celsius. In our daily lives, 40°C/104°F (313 K) is a rather hot day and 100°C/212°F/373 K) is about

The speed of light is 299,792,458 metres per second (670 million miles per hour)

0.000000000000000000000000000000000000000001 (10^{-42}) seconds

The cosmic horizon grows ever larger as time goes by, but never fast enough to allow us to see what's beyond it.

the temperature of the tea that burns your tongue. At this moment, 10^{-43} seconds after Time Zero, the temperature of the universe is 10^{32} K – one million million million million million times hotter than that boiling tea.

Luckily for all of us, the temperature drops rapidly as space expands. Indeed, within 10^{-35} seconds, the temperature is already a thousandth of a per cent of what it was after a Planck time. Even so, there is no way particles of matter can be produced at or survive such temperatures. So the contents of the universe right now consist solely of light – energy in the form of electromagnetic waves and photons.

The intensity and dynamism of that light is almost unimaginable today. Yet in the midst of that sub-microscopic maelstrom, a fundamental physical quantity is established: the speed of light is set at 299,792,458 metres per second (670 million miles per hour). Why that speed exactly? We still don't know. Yet this velocity is so basic to the universe and its future that we now measure length with it. The length of 1 metre (3¼ feet) is officially defined as the distance light travels in a perfect vacuum in 1/299,792,458 second.

By human standards, the speed of light is blindingly fast – a beam of light could travel from New York to London and back more than 25 times in a single second. In our vast universe, however, getting from star to star at the speed of light takes years. Even in the wake of the Big Bang, just after the Planck time, the expansion of the universe is starting to carry parts of the cosmos far enough away from us that we will never be able to see light from there again. A boundary around our observation of the cosmos is beginning to form, a function of the age of the universe multiplied by the speed of light – an ever-growing cosmic horizon.

The Big Bang – How it Happened

The Big Bang is awesome – maybe even a little overwhelming. How did the birth of the universe all happen?

Our curiosity about our cosmic origin of course predates our ability to study it. Long before humans began using science to solve the mysteries of nature, we were already telling creation stories, looking around us for clues and using our imaginations to extend our inferences to the very beginning of everything. Supernatural or divine explanations involving animals, spirits and gods came from every early human culture.

Probably the most common thread that ran through all these creation stories was the change from stillness and darkness into activity and light. In scientific terms, this means that energy had to be added into the universe where previously there had been none. Given the conditions of the beginning of time, two scientific ways to add the necessary energy have been proposed: quantum fluctuations and symmetry breaking.

Intense energy concentrated in a tiny volume bursts forth as light

In quantum theory, it is possible to have a 'bubble' of energy appear spontaneously, but then disappear so quickly that the energy surge goes unnoticed. These quantum fluctuations can be enormous if the amount of time they last is tiny. One possible energy source for the Big Bang might thus be a quantum fluctuation that grew out of control – instead of vanishing, the energy ballooned outwards and drove the expansion of space and time from a single point into an entire universe.

The ancient Greek philosopher Aristotle (384–322 BCE) suggested that at the beginning of time, some sort of 'prime mover' was responsible for starting the clockwork of the universe; all of history has unfolded since that first motion.

0. 00000000000000000000000000000000000000001 (10^{-40}) seconds

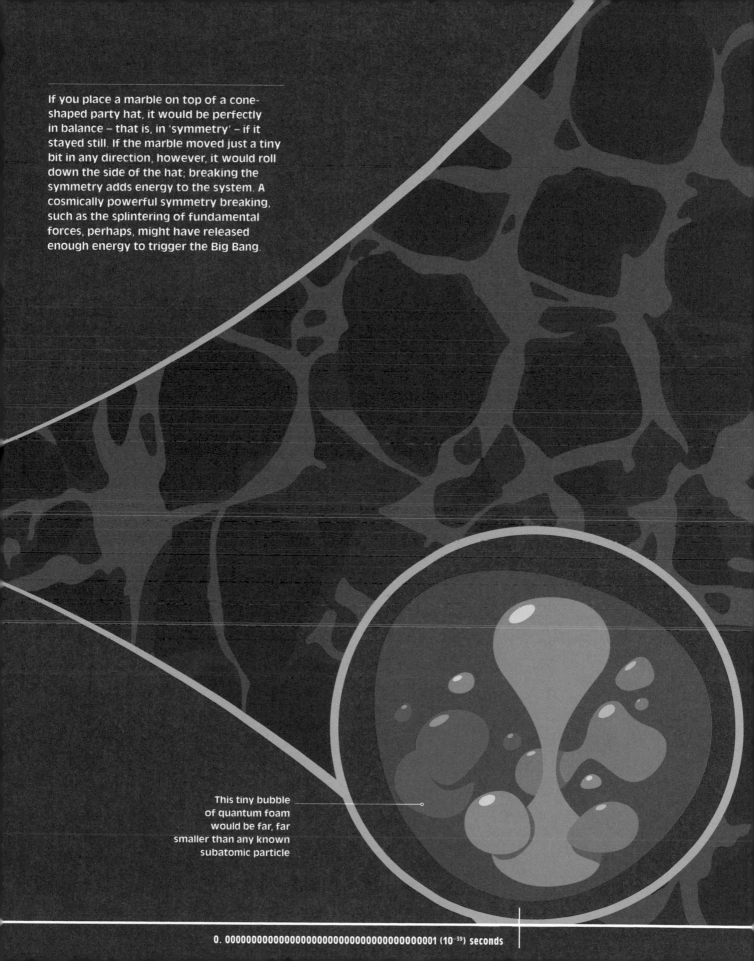

If you place a marble on top of a cone-shaped party hat, it would be perfectly in balance – that is, in 'symmetry' – if it stayed still. If the marble moved just a tiny bit in any direction, however, it would roll down the side of the hat; breaking the symmetry adds energy to the system. A cosmically powerful symmetry breaking, such as the splintering of fundamental forces, perhaps, might have released enough energy to trigger the Big Bang.

This tiny bubble of quantum foam would be far, far smaller than any known subatomic particle

0. 000000000000000000000000000000000000001 (10^{-39}) seconds

M Theory

To find the origin of the cosmos, we may need to look beyond length, width, height and even time.

Whether symmetry breaking, quantum fluctuation or some other mechanism launched the expansion of the universe, the Big Bang definitely required a whole lot of energy. Did it all come from ... nothing? More than that, the universe is expanding into ... what?

Just as humans have always wondered what lay beyond the mountain or over the ocean, scientists have investigated the possibility that our universe is one entity within an even larger structure. One thing we know is that space isn't expanding into more space, and there wasn't a time before the Big Bang – all of the space and time that exists came into being at that moment. We're not looking for extensions of past and present, nor of length, width and height – rather, we're examining the possibility that entirely different dimensions exist beyond our own.

Imagine, for example, two flags flapping in the wind next to one another. Every once in a while they'll touch briefly; that point of contact is connected to both flags, yet is not exactly part of either flag. Now take that analogy to a much higher level: if two undulating five-dimensional 'membranes' were to touch, might they produce a single four-dimensional spot? And if three of those four dimensions were length, width and height, the fourth dimension produced by this contact could be time – the time that started to run at the moment called the Big Bang.

If symmetry breaking was indeed important in powering the universe at the beginning of time, then the universe and its subatomic constituents are likely to exhibit supersymmetry – a theoretical structuring of matter and energy at the most fundamental levels. In a supersymmetric universe, every kind of particle has a partner called a sparticle – and the unseen interactions between them are a key part of all of the physical phenomena we observe. Supersymmetry could be confirmed if sparticles are detected by high-energy physics facilities such as the Large Hadron Collider at CERN.

Part of M theory is the idea that structures with more than four dimensions interact beyond our universe. This is the culmination of many versions of what physicists call string theory, and is mathematically very elegant. Unfortunately, it provides no feasible way to test its predictions with observations or experiments. Physicists jokingly say that the 'M' could stand for membrane, mystery or even magic.

Inflation

Just after the Planck time, the speed of light was fast enough to transfer energy from one end of the then-tiny universe to the other.

This maintained uniformity and equilibrium throughout the cosmos. As the universe expanded, however, vast differences should have built up as distances grew ever larger; and by the present day, we would see those differences in the distribution of matter and energy across the cosmos. That's not what astronomers observe today; indeed, at large scales the universe looks just about the same in every direction, and its geometric perfection is uncannily exact. We call these puzzles the horizon problem and the flatness problem.

One way to resolve these two problems is to have a period of time when the universe didn't just expand linearly, but rather inflated exponentially. Just before the different regions of the universe lost contact with one another, a sudden and wild inflation occurred that stretched space by at least a trillion trillion times in every direction. When normal expansion resumed, each region had the old, uniform pre-inflationary conditions imprinted in them, so naturally every volume would then grow from the same starting point and eventually look like every other similarly sized volume.

For inflation to work, it had to start by around 10^{-35} seconds after the Big Bang and end by about 10^{-32} seconds. That's not much time at all by human standards, but it's more than 10 billion Planck times – and more than enough to stretch the cosmos into the shape we see today, setting us off on the inexorable expansion that has continued to the present day.

As with the Big Bang itself, a titanic influx of energy would have been needed to power the inflationary period. One proposed source of this energy is another symmetry breaking – possibly when one fundamental force 'broke' into two, creating the electromagnetic force and the nuclear force.

While the idea of an inflationary period resolves important issues about the birth and growth of the cosmos, it also creates an unintended consequence. What prevents inflation from happening repeatedly in different parts of a four-dimensional space-time like our own? The denizens of this space-time could never escape their own inflated regions of space, and the result could be a vast number of expanding mini-universes, embedded together yet forever separated by expansion – creating, in effect, a multiverse.

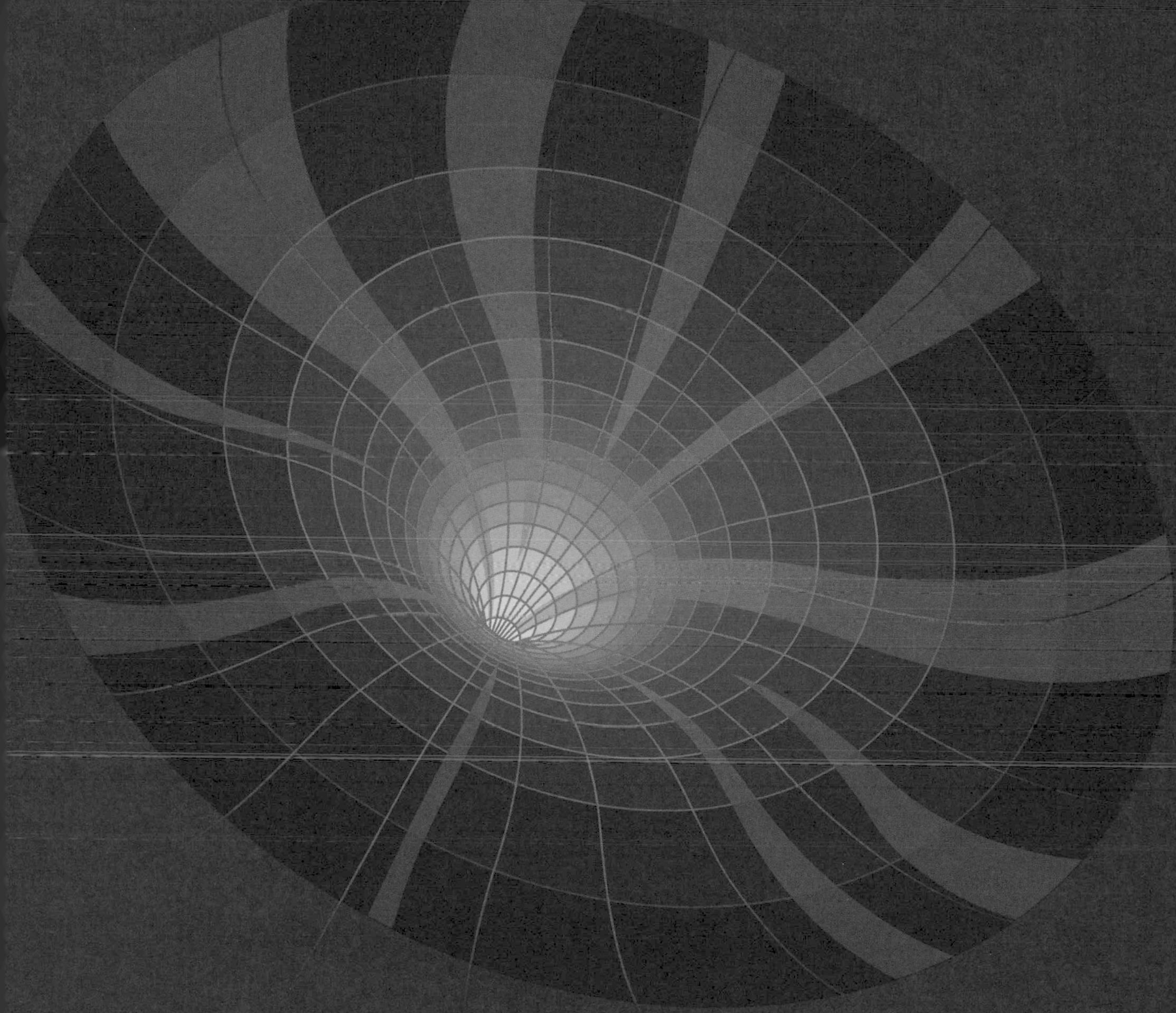

0.00000000000000000000000000000001 (10^{-32}) seconds

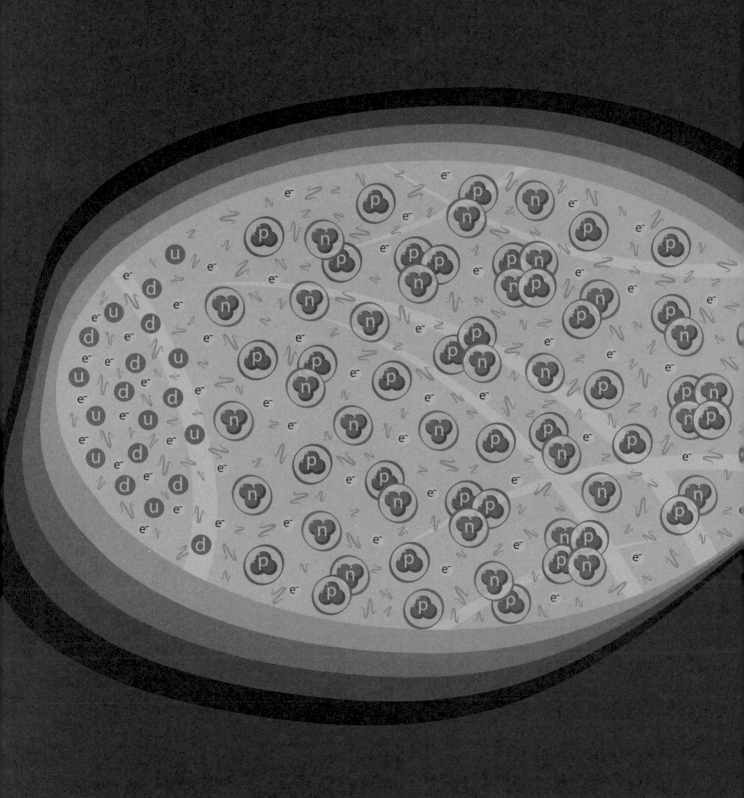

0.0000000000000000000000000000000001 (10^{-32}) seconds

2

Five Minutes After the Big Bang

The Stuff of the Universe is Produced

The infant universe is growing, glowing and scorching hot.

A quadrillionth of a second after the Big Bang, the temperature of space is more than a billion billion degrees Celsius. This early in its history, the cosmos can contain only one thing: energy. Nothing else can form or take shape in the blistering heat.

Happily, this inhospitable environmental condition doesn't last for too much longer. Thanks to an amazing process called 'electroweak symmetry breaking' – or, more simply, the Higgs mechanism, named after one of its discoverers (see pages 30–31) – particles of energy soon start to turn into particles with mass, forming the first bits of the matter that we're made of today.

How does this seemingly magical transformation happen? Imagine standing by the seashore on a humid day. If a cold wind suddenly blows towards you from the sea, the water vapour in the air will condense into droplets and a fog will roll in. The basic substance of the droplets is the same as that of the vapour; it has, however, changed the way it's organized – in this case, from gaseous to liquid form. Energy and matter relate to one another in a similar way: as Albert Einstein showed with his famous equation $E=mc^2$, energy and matter are two sides of the same coin, just as pounds and dollars are both currency, connected by a straightforward conversion factor.

Early in the life of the cosmos, there is no cool ocean breeze. Instead, space expands continuously as time marches on, so the blazing-hot energy spreads out and the temperature drops in response. After the universe ages another thousandfold, to the ripe old age of a trillionth of a second, the temperature has dropped to just a few hundred trillion degrees. Although that still seems unimaginably hot to us, it crosses a critical cosmic threshold, triggering the Higgs mechanism.

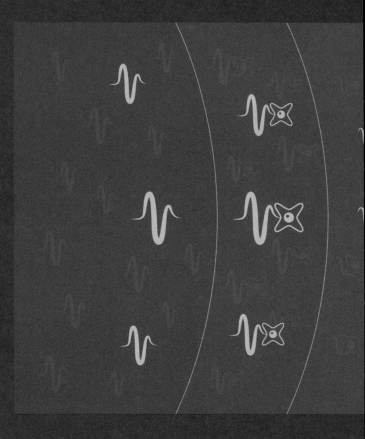

Once massive particles start appearing, they proliferate in both number and variety. With names like bosons, fermions, quarks and leptons, they start swimming around in a dense cosmic soup. As space grows still larger and cools still further, the soup becomes a zoo of particles. Many of them then start to combine into even more massive particles. Some particles, meanwhile, remain massless – such as photons, which we can sense as light, and gluons, which

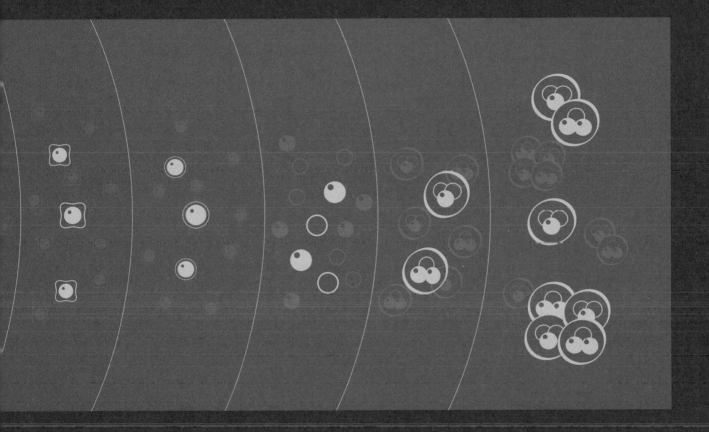

bind the quarks. Together, quarks and gluons assemble into protons and neutrons, which in turn combine to make deuterium and helium. When all is said and done, the universe is filled with atomic nuclei.

All this particle creation and nucleosynthesis feels like it should take aeons of cosmic time – but by human reckoning, it happens in just a few brief minutes. Even more surprisingly, the newly made matter isn't all immediately annihilated by antimatter and turned back into energy, even though there is ample opportunity for this to happen. Against the odds, for reasons still scientifically unknown, a tiny fraction of the matter sticks around. These lingering remains will persist as time marches on; and that's a very good thing, because these are the particles that will become the building blocks of the stars.

Four Forces in Place

Before matter can exist, fundamental forces take hold of the cosmos.

After the inflationary period, the universe returned to expansion at its original rate for a long time, all the way until about 100 million trillion times its age at the end of inflation, growing and cooling without incident. Then, at the ripe old age of roughly one-trillionth (10^{-12}) of a second, a new cataclysm shook the cosmos. Why?

One likely explanation is that a fundamental symmetry breaking occurred. In that model, there were at least two previous symmetry breaking events: the first happened at the Planck time, around 10^{-43} seconds, and the second happened at the start of inflation around 10^{-35} seconds. At each of these breaks, the forces of the universe splintered; the first break split the original unified force into two, and the second break turned two into three.

Now, at 10^{-12} seconds, this third break turned three into four: gravity, the strong nuclear force and the electroweak force became gravity, the strong nuclear force, electromagnetism and the weak nuclear force. Just like the other two force-splitting events, we don't know exactly why this happened; it could be that the average temperature of the universe had dropped to just the right level – below about 1 million billion degrees – to allow the symmetry breaking. In any case, with the four fundamental forces now in place, the emergence of massive particles in the universe could begin – the most basic building blocks of matter that we can touch and see today.

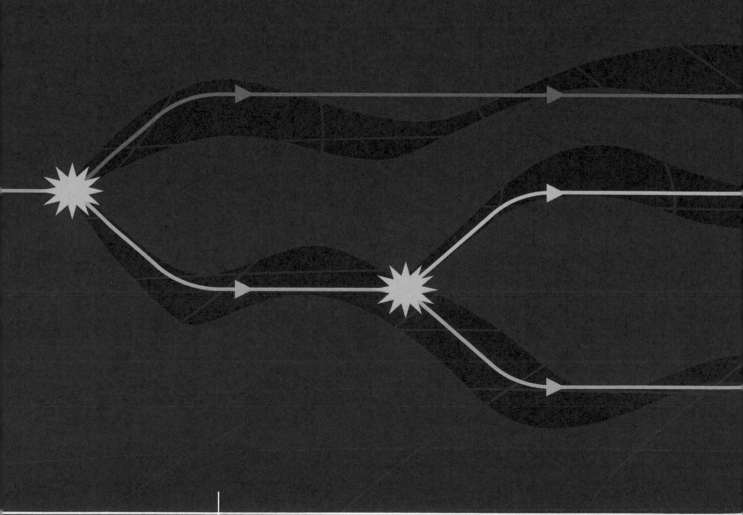

In common parlance, a force is a push or a pull. Cosmically speaking, the four forces are the fundamental ways in which pushes and pulls can be created in the universe. Each of these forces stems from particles sensitive to that force interacting with a 'potential field' – a region of space where the force has influence. These days, for example, a little magnet creates a potential field that interacts with the metallic particles on your refrigerator door to create a pulling electromagnetic force between the magnet and the door. The magnet will not stick to a tree, though, because wood particles do not interact with an electromagnetic potential field. Back in the infant universe, of course, there were no magnets, doors or trees; one-trillionth of a second after the Big Bang, there were just energy and the forces. The first particles that formed in the universe had to serve a key function.

Among the four fundamental forces, the strong and weak nuclear forces have potential fields that extend only to subatomic scales, while electromagnetism and gravity can reach across planetary or galactic scales. In terms of particle-for-particle strength, the strong nuclear force is about a thousand times more powerful than the electromagnetic force, which in turn is more than a trillion times more powerful than the weak nuclear force. With far less than one-trillionth the strength of the weak nuclear force, gravity is by far the weakest force on this scale.

Gravity

Strong nuclear force

Weak nuclear force

Electromagnetism

$0.000000000001 \ (10^{-12}) \ \text{second}$

Subatomic Particle Soup

Matter first emerges in the universe as a brothy quark–gluon plasma.

If current scientific theories about the universe are correct then starting around this time, one-trillionth of a second after the Big Bang, each of the four forces start being carried from one part of the universe to another by its own kind of particle called a boson. Gravity is being carried by gravitons, the strong force by gluons, electromagnetism by photons and the weak force by W and Z particles. Just around the time the electroweak force splits, though, another potential field comes into prominence: the Higgs field. This field interacts with the W and Z particles, creating a drag on their motion the way water in a pool drags on a swimmer. That drag makes the particles heavy – in other words, for the first time in cosmic history, objects have mass.

These relative properties of the four forces are at the forefront of the evolution of the cosmos before the universe is one-millionth of one second old. The first massive pieces of matter begin to appear now, as the universe has expanded and cooled enough for them to form. There are six kinds of these particles, known as quarks, and they are endowed with mass as they interact with the Higgs field. They also have a bit of electric charge – three of the six have one-third the charge of a present-day electron, and the other three have two-thirds the charge of a present-day proton. There are antiquarks, too – antimatter versions of the quarks with opposite charges. The universe is still so small that the strong nuclear force controls the quarks' motion and behaviour. Gluons – and there are eight kinds of them – swirl and flow all around the quarks. The result is an electrically charged, viscous brew of astounding energy and density.

Photons are particles of light and ubiquitous in the universe today. Using powerful particle accelerators, scientists have detected gluons, W and Z particles, and Higgs bosons. All of these different particle types seem to follow a set of physical rules consistent with the fundamental force and mass picture described above. The remaining sticking point is the graviton, which has yet to be detected experimentally; indirect evidence of gravitons may exist, however, in studies of gravitational waves, produced by distant collisions of black holes that have been observed on Earth.

BIOGRAPHY

Satyendra Nath Bose (1894–1974) was an Indian theoretical physicist and mathematician who, together with Albert Einstein, discovered the properties of subatomic particles that could carry force. Today we call those particles bosons in his honour. Bosons can congregate with similar particles to produce Bose–Einstein condensates with amazing material properties.

In 1964, British physicists **Peter Higgs** (b. 1929) and **Tom Kibble** (1932–2016), Belgian physicists **François Englert** (b. 1932) and **Robert Brout** (1928–2011), and American physicists **C. R. Hagen** (b. 1937) and **Gerald Guralnik** (1936–2014) proposed theories about the existence of a particle and its corresponding potential field that would explain why particles have mass. Their theories were proven correct nearly half a century later, when on 4 July 2012 the experimental detection of the Higgs boson was announced.

0.00000001 (10⁻⁸) seconds

0.0000001 (10⁻⁷) seconds

Subatomic Particle Zoo

In one-thousandth of a second, a goopy plasma becomes a plethora of particles.

Photons and gluons, unlike their boson counterparts W-, W+ and Z, do not interact with the Higgs field and thus have no mass. They still carry energy, though, and at the very high temperatures present during the first moments of the universe they can be so energetic that they transform into massive particles.

In the quark–gluon plasma, this happens continuously as the massless gluons interact with the massive quarks, integrating and disintegrating as various combinations are built and pulled apart in the swirling heat. As the plasma cools further, dropping to a temperature of many trillions of degrees, the quarks and gluons finally start coalescing into larger composite particles called hadrons. The six kinds of quarks – with the cheerful names up, down, charm, strange, top and bottom – combine in quark–antiquark pairs to make mesons, or in three-quark trios to make baryons.

Two kinds of quarks eventually become stars of the show: up, with a charge of positive two-thirds, and down, with a negative one-third charge. Two up quarks and one down quark, knit together with gluons, make positively charged protons, while one up quark and two down quarks, also bound by gluons, construct electrically neutral neutrons.

One ten-thousandth of a second after the Big Bang, the cosmic background temperature is down to a few trillion (10^{12}) degrees, and the gluon-and-quark dance has now started producing hadrons. Now another group of particles, called leptons, make their entrance on to the cosmic stage. These are the lightest of the elementary particles that contain mass. The best-known of the six leptons is the electron, a constituent of every molecule in the universe today and the carrier of negative electric charge that we harness in our technology every day. The muon and tauon also carry negative electric charge. Each of these three charged leptons has a partner with no charge and only a tiny amount of mass; they are the three types – or 'flavours' – of neutrinos, ghostly particles that can travel unhindered through just about any amount of material in the universe.

The key to turning energy into matter and back is $E=mc^2$, perhaps the most famous equation in science. Derived by Albert Einstein in 1905, this formula explains how much energy (E) is needed to produce a particle of mass (m). The conversion factor is the speed of light squared – an enormous number; the amount of energy used in the entire United Kingdom each year, converted fully into mass, would yield barely 100 grams (3½ ounces) of matter.

The menagerie of basic subatomic particles fits into a theoretical framework known as the Standard Model. Three pairs of quarks – up and down, charm and strange, top and bottom – share properties with one another, as do three pairs of leptons – electron and electron neutrino, muon and mu-neutrino, tauon and tau-neutrino; these dozen particles are the fermions. Each of these particles has a matching antiparticle – the antimatter version of itself. Another set of particles – eight kinds of gluons, W+, W-, Z, photons, gravitons and the Higgs particle – are the bosons.

BIOGRAPHY

Albert Einstein (1879–1955) was a graduate student in Switzerland in 1905, working as a clerk in a patent office to support himself and his family. That year, he published four ground-breaking papers describing the theoretical underpinnings of much of modern physics. A decade later, Einstein discovered the general theory of relativity, which mathematically explains how space, time and gravity tie together in the shape and structure of the universe.

QUARKS

LEPTONS

Up

Charm

Top

Electron

Muon

Tau

Down

Strange

Bottom

Electron
neutrino

Muon
neutrino

Tau
neutrino

BOSONS

Gluon
(8 types)

Photon

Higgs boson

Graviton (not yet
directly detected)

Z boson

W boson
(2 types)

Nucleosynthesis Begins

When energy converts into matter, it always makes particles two by two – that is, as a particle–antiparticle pair.

The individuals in the pair are equal, but their opposite natures are such that if they ever touch again they will annihilate one another, and both will become energy once again.

One of the best-known processes of this type is simply called pair production: a high-energy photon, influenced by a nearby particle, spontaneously turns into an electron and an anti-electron (also called a positron). From about one-thousandth of a second to one-hundredth of a second after the Big Bang, pair production is fast and furious, and electron–positron annihilation is equally so. Other particle pairs, such as protons and antiprotons, are also being created and destroyed in the bubbling soup of mass and energy contained in space.

The universe keeps expanding, though. That means the same amount of heat energy is spread out into a larger and larger volume, so the average temperature in space keeps dropping, as does the average energy of the photons flying through space. About one second after the Big Bang, neutrinos are able to travel through the universe without being destroyed. Seconds later, protons and neutrons are produced that are not immediately destroyed, but stay around long enough to interact. Eventually, there are about seven protons in the universe for every neutron.

At this point, when a single proton interacts with a single neutron, the result is almost always a benign collision – they exchange some energy with one another, but quickly part ways essentially unchanged, sort of like colliding marbles or snooker balls. But once the temperature of the universe drops to about a few billion degrees, a new and amazing physical reaction begins to occur. Every once in a rare while, a proton–neutron collision results in a merger – and the two particles form a new, composite particle called a deuteron. This is the first kind of multiparticle atomic nucleus to be made, and marks the start of Big Bang nucleosynthesis.

Different versions of elements whose nuclei have the same number of protons but different numbers of neutrons are called isotopes. A proton is the nucleus for a typical hydrogen atom; a deuteron is the nucleus for the isotope of hydrogen called deuterium. A third isotope of hydrogen exists as well that has one proton and two neutrons – it's called tritium.

The modern picture of an atom is that of a small, dense nucleus of matter surrounded by a cloud of orbiting electrons. The positively charged nucleus balances out the negatively charged electron cloud, resulting in a neutral atom with no net charge. At the high temperatures of the early universe, however, an electron cannot stay bound to a nucleus because there are too many high-energy collisions with all the surrounding subatomic particles.

Oddly, the mass of a proton or neutron – a property it gains thanks to the Higgs field – is far greater than the combined mass of the three quarks that are its constituents. The vast majority of the mass in atoms and molecules today comes from the energy of the gluons, which are massless, that bind each proton or neutron together. The three 'valence quarks' are surrounded in their particle by a 'sea' of quark–antiquark pairs that come and go as the gluons continually form and split. It's an amazing example of Albert Einstein's signature formula that shows the equivalency of energy and mass, $E=mc^2$.

The photon is the massless boson that carries the electromagnetic force. In that role, photons serve double duty as quantum mechanical units of energy that travel through the universe as both a particle and a wave simultaneously. A photon's energy determines its wavelength and its colour; a wave of blue light, for example, carries more energetic photons than a wave of red light. X-rays and radio waves are examples of photons with wavelengths beyond what human eyes can see.

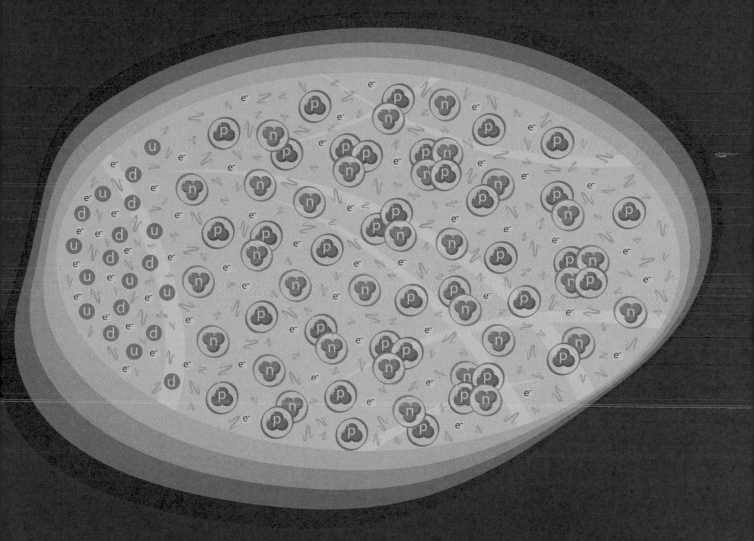

At first, matter and antimatter particles
are crowded together, annihilating
and reforming. As space expands and
temperature drops, antimatter disappears.
Eventually, the matter particles that
remain are mostly neutrinos, electrons,
protons and neutrons.

0.1 (10⁻¹) seconds

1 (10⁰) second

Nucleosynthesis – Heavier Nuclei

Deuterons become increasingly plentiful as nucleosynthesis progresses, reaching their peak production in less than two minutes.

As they multiply, new processes spring into action. A deuteron can combine with a free neutron to form a new nucleus with one proton and two neutrons – a 'hydrogen-3' – or with a free proton to form a different nucleus with two protons and one neutron – a 'helium-3'. Deuterons combine with one another to form new nuclei – either with one proton and two neutrons, with one proton escaping, or with one neutron and two protons, with one neutron flying free.

The peak of Big Bang nucleosynthesis arrives as deuterons, hydrogen-3 and helium-3 interact willy-nilly, producing a wide variety of combinations. Ultimately, the net effect is that just about every pair of deuterons go into forming a nucleus with two protons and two neutrons: a 'helium-4'. Five minutes after the Big Bang, more than three-quarters of the mass of the universe resides in protons, and just under one-quarter of the mass in the form of helium-4 nuclei. The tiny bit of remaining mass is in the form of leptons and a residue of neutrons, deuterons and other nuclei.

It may seem odd that the nuclei of only two elements, hydrogen and helium, have been produced in the entire history of the universe so far. What about carbon, oxygen, iron, silver, gold and all the other elements that make up everything in the world? Their nucleosynthesis would have to wait – for a very long time.

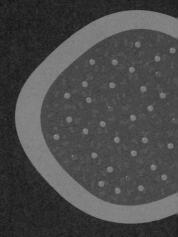

The modern periodic table of elements uses a symbolic notation system for atomic nuclei with superscripts showing the number of baryons: a proton is ^{1}H, a deuteron is ^{2}H, hydrogen-3 is ^{3}H, and helium-3 and helium-4 are ^{3}He and ^{4}He respectively.

At this point in cosmic history, the leftover nuclei other than ^{1}H and ^{4}He comprise less than one-tenth of 1 per cent of the mass in the universe. A very tiny number of these heavy nuclei survive to this day in the form of lithium-7 (^{7}Li), ^{2}H and ^{3}He; astronomers study the proportions of those early elements to deduce the detailed conditions of the universe in those few minutes after the Big Bang.

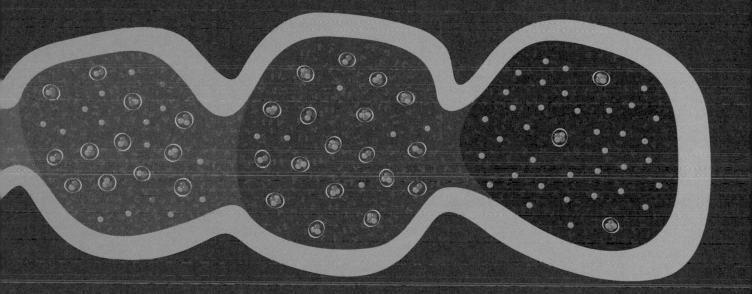

Protons (blue) and neutrons (red)
combine in a variety of ways for several
minutes, until ultimately settling down
almost exclusively into hydrogen and
helium nuclei. Hydrogen now dominates
all the luminous matter in the universe.

3 minutes

4 minutes

5 minutes

Matter–antimatter Imbalance

We are here now – that is, the particles that will eventually make us.

Less than 10 minutes after the Big Bang, the universe has expanded to a size so large that the average energy density in space is no longer high enough to induce nucleosynthesis. Every neutron not bound into a nucleus lasts on average less than 15 minutes before it spontaneously turns into a proton, an electron and an antineutrino. The production of deuterium and helium stops, and within a few hours the number of protons stabilize. Until those high-energy conditions reappear again, the amount and ratio of the different elements and isotopes has stopped changing for now.

But wait. Why are there elements and isotopes at all? When energy produces matter, such as in pair production, equal amounts of matter and antimatter are created. Within a minute after the Big Bang, all electrons and positrons moving freely around the universe should have collided and destroyed one another. As nucleosynthesis happens, there should be as many particles constructed of antimatter as there are of matter. As they continue to move randomly around the universe and come into contact with one another, they should annihilate; and slowly but surely, all the matter should be gone. Yet matter still exists.

Clearly, there must be some sort of imbalance that tilts the cosmic pitch in favour of matter. That's a good thing, because otherwise we could not exist today. It means, though, that a basic symmetry of the universe must have been broken at some point to cause the tilt. Was it a side effect of one of the forces freezing out? Or has some other kind of asymmetry been built into the fundamental nature of the particles of the universe?

Currently, one of the most promising hypotheses explaining the matter–antimatter imbalance involves a combination of both those ideas, via a category of so-called grand unified theories (GUTs) that try to explain how the strong nuclear, weak nuclear and electromagnetic forces interacted when they were unified, and what happened after they split apart. If some version of a GUT is correct, then there is a straightforward explanation of why a tiny percentage more matter was created in the early universe than antimatter.

With this explanation, however, comes a consequence: all complex particles of matter must eventually decay into simpler, less massive particles. The ultimate test is to see if protons decay – and despite decades of experiments, not a single proton decay has ever been measured. The mystery of matter's very existence in the universe remains unsolved.

One way to visualize the matter–antimatter imbalance in the early universe is to think of a football pitch in which 10 billion pennies are all balanced on their edges. Now knock them all over all at once; and even though every penny has an equal chance of falling heads or tails, exactly 5,000,000,001 of them land on heads and 4,999,999,999 of them land on tails. Now imagine if that were to happen over and over, every single time the pennies are reset and knocked over again, for countless iterations. What could possibly cause this unwavering discrepancy?

One of the most significant experiments in particle physics today is the Super-Kamiokande facility, located a full kilometre underground in the Mozumi Mine near Hida, Japan. A tank of 50,000 tons of highly purified water is surrounded by a vast array of photosensors; it is ideally suited to detect neutrinos passing through Earth as they fly through space, as well as the decay of any proton that occurs in the water. So far, plenty of neutrinos have been observed, but no proton decays. As of today, this means that the average lifetime of a proton must exceed 10^{34} (10,000,000,000,000,000, 000,000,000,000,000,000) years.

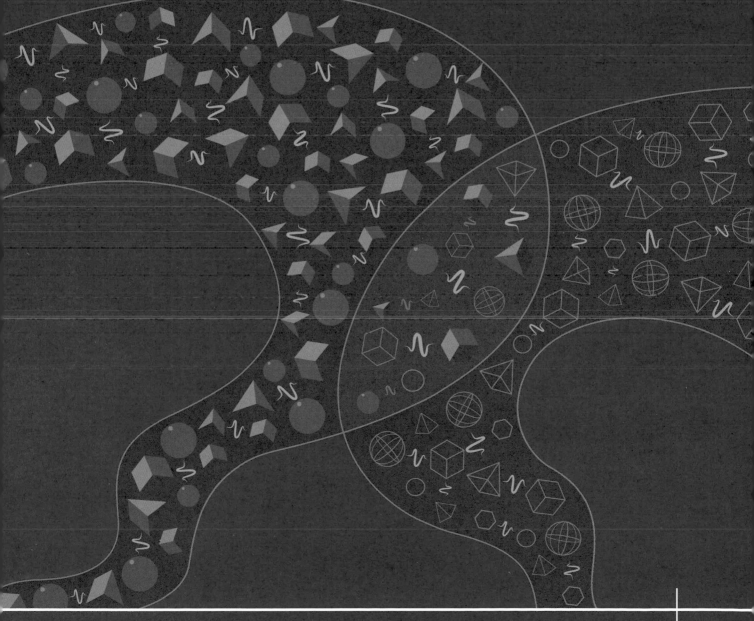

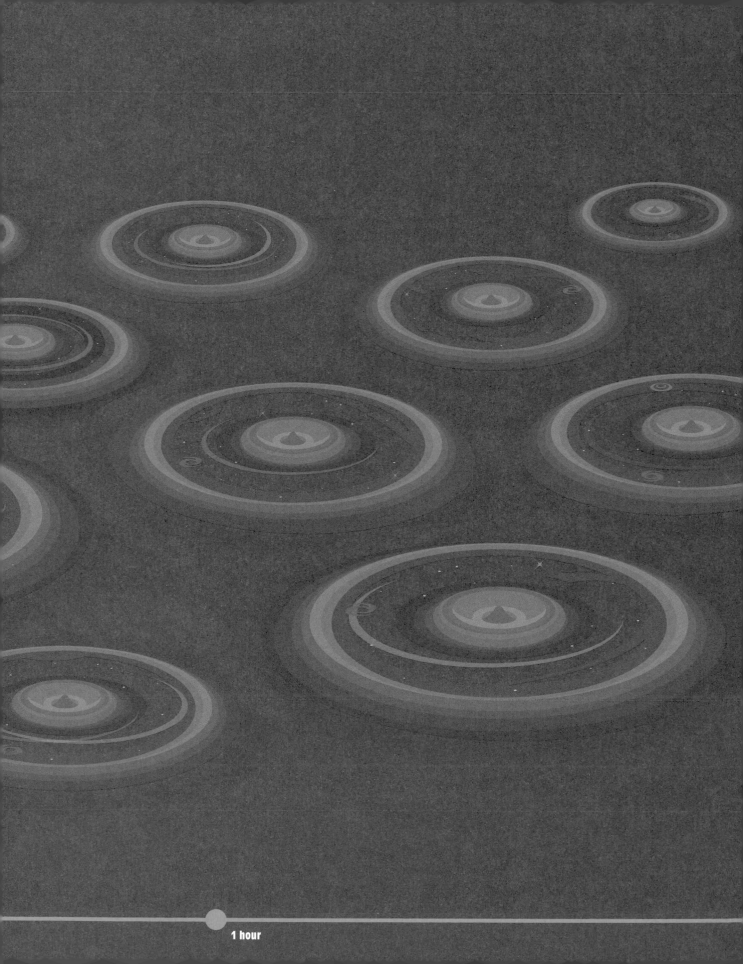

3

400,000 Years After the Big Bang

The Large-scale Structure of the Universe is Established

Today's universe has structures on scales large and small.

The tiniest grain of sand on a terrestrial beach contains atoms and molecules bonded together by nuclear and electromagnetic forces – yet those same forces hold the particles apart. The result is a solid structure that is held rigid, able to move as a unit freely within, while clearly distinct from the water and air that surround it. On a cosmic scale, on the other hand, structures tend to be held together mostly by gravity; so every grain of sand, drop of water and puff of air is held close to Earth's surface – and in their multitudes, different from each other as they are, they comprise the planet on which we live.

Minutes after the Big Bang, the universe has no structure – just atomic nuclei, subatomic particles and vast amounts of energy coursing around in a jumbled mess. Somehow, all of that chaos must have spawned the structure we observe at present – planets, stars, galaxies and how they're distributed throughout space. What happened?

Once again, cosmic expansion drives the change. As the volume of the universe grows, and minutes turn to hours, days, weeks and months, the temperature of space continues to drop from the scorching levels that existed when Big Bang nucleosynthesis ended. The speed of the expansion – that is, how big the universe is now compared to some previous moment – is now a way to measure the passage of time itself. And that time continues to pass, into years, decades, centuries and millennia – until the temperature drops from millions to just thousands of degrees.

While on average every spot in space is still hotter than a blast furnace, matter and energy can begin interacting more systematically rather than just randomly. Imagine pouring a drop of milk into a cup of tea: if the beverage is boiling hot, the milk mixes immediately with the tea, but if it's a little cooler, the milk flows in curlicues throughout the tea for a while, creating a fleeting, swirling structure. If the tea gets rapidly colder, the structures won't dissipate evenly into the tea; the swirls will remain, and you'll have to use a spoon to stir it all into a uniform mixture.

In the universe, of course, there is no cosmic spoon to mix everything back up, so structures that form will stay around and keep growing as space cools down. On the other hand, there is no milk-filled cosmic creamer either; structure must arise from whatever matter already exists, not from some outside source. In much the same way that protons and neutrons congealed out of quark–gluon plasma moments after the Big Bang, much larger structures do that now – the hydrogen and helium nuclei bond together with electrons to form atoms. Photons – those massless pieces of light that carry the electromagnetic force – stop dragging on matter, and begin travelling through space independently. Whatever structures that exist now, subtle as they may be, are the seeds of the vast variety of amazing cosmic phenomena to come.

Redshift

Light travelling through the universe is stretched by expanding space, providing a physical way to measure cosmic time.

If you haven't ever done this before, draw a wavy line on a rubber band and then stretch the band out. Notice that the wave stretches – and, of course, the distance between the peaks of the wave increases. Congratulations! You have just demonstrated how astronomers keep cosmic time: redshift.

Space-time – the fabric within which all of the stuff of the universe is embedded – is elastic and stretchy, just like a rubber band, except in four dimensions across vast distances. Your rubber band stretches in one dimension (length), while the surface of a balloon expands in two dimensions as it is inflated. For three-dimensional expansion, perhaps you could imagine a puffy loaf of raisin bread; as it bakes, the dough expands, carrying the raisins further and further apart.

The amount of space-time we inhabit on Earth today is so small compared to the universe as a whole that we can't perceive either its expansion or its curvature in our regular lives. A little beetle walking on a giant balloon, for example, can't tell it's on a curved surface until it's moved a fair fraction of the way around the circumference. Similarly, we have to look far into space to see any sign of the cosmos stretching.

Here's where the wavy line comes in. Photons travelling across the universe have a wavelength, and that wavelength grows as the universe grows. Shorter wavelengths of visible light look bluer to our eyes, whereas longer wavelengths of light look redder. So, when a beam of light emitted a long time ago reaches Earth, we will observe it after its wavelength has been stretched, and its colour will have been shifted towards redder colours – in other words, it's redshifted. The higher the redshift, the more time has passed since it was emitted.

With redshift, we let the expansion of the universe mark the passage of time. We don't use a pocket watch or a pendulum – the cosmos itself is our clock.

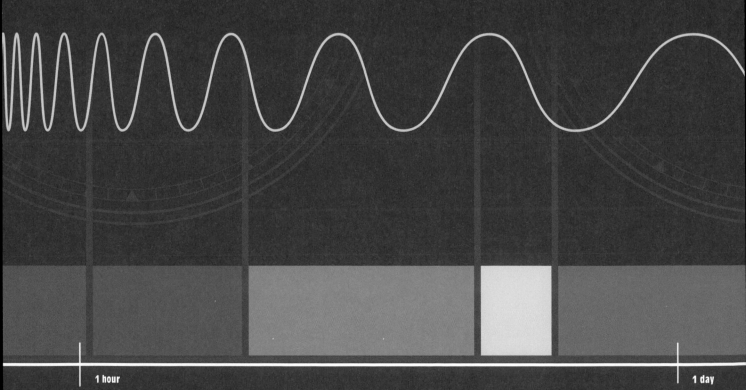

1 hour

1 day

Light we can see with our eyes comprises a small fraction of all the light that flows through the universe. Light invisible to us is often grouped into categories based on its wavelengths: gamma rays, X-rays, ultraviolet, infrared, microwaves and radio waves. All light together is known as the electromagnetic spectrum.

As the age of the universe grows beyond minutes into years and centuries, distances across the universe grow large as well. The distance light travels in one Earth year is a unit of length called a light year, and is equal to about 9,460,000,000,000 (9.46 trillion) kilometres, or about 5.88 trillion miles. For comparison, the distance from London to Hong Kong is 0.000000001 light year. An even longer unit of length is the parsec, which is equal to about 3.26 light years.

BIOGRAPHY

The Reverend Monsignor Georges Henri Joseph Édouard Lemaître (1894–1966) was an astronomer, mathematician and physics professor who was the first person to propose the 'hypothesis of the primeval atom', which would later become known as the Big Bang. He connected the observed redshift of distant galaxies with the expansion of the universe, and in 1927 published the first measurement of the rate of cosmic expansion.

The amount of redshift observed in a beam of light indicates how long ago the light was emitted. If the observed wavelength of light is twice that of the light's original colour, then that light left its source when the observable universe was one-half its original diameter. If it's three times the wavelength, it left when the observable universe was one-third its original diameter; four times the wavelength, one-quarter the diameter; and so on.

Baryon Acoustic Oscillations

Ripples in the cosmos are created by the 'ringing' of the universe as matter and energy flow together through space-time.

The tremendous power of the Big Bang rang the infant cosmos like a well-struck gong. As the universe expanded and grew, other vastly energetic events kept space and time ringing with sound-like waves. Amazingly, astronomers can see the remnants of those waves even today. How?

If you drop a pebble into a pond, a circular wave emanates from the point of impact outwards across the surface of the water. Now drop in a handful of pebbles; each pebble's wave spreads and overlaps with the wave of every other pebble, and soon there is a jumbled pattern of crests and troughs. Although it may seem that the pattern is an indecipherable mess, it is mathematically possible for scientists to reconstruct and calculate how and when the pebbles originally hit the water – for a time, anyway, before the waves all fade away. So imagine what would happen if, moments after the pebbles struck the pond's surface, the pond froze solid. The pattern would be preserved in the ice, so even if the pebbles have sunk out of sight, we could figure out how the original splash occurred.

This Earthly scenario is analogous to what happened in the expanding early universe. As the temperature dropped, the patterns of vibrations froze into the matter – specifically, baryon particles such as protons and neutrons – spreading throughout space. The size of these baryon acoustic oscillations increased as the universe grew, peaking at a maximum when the oscillations froze into place. Eventually, atoms and molecules formed closer together in the rings and shells where the baryons were bunched together; and today, the positions of galaxies and stars trace out those patterns as well.

Baryon acoustic oscillations, as their name implies, are a kind of sound wave. Unlike a light wave, which can travel through a vacuum, these waves must pulse through a medium the same way that a bird's song moves through air or ocean waves move through water. Early in the history of the universe, oscillations could travel through the soup of particles and energy that occupied space; they can no longer do so today, because matter is too thinly spread and space is too expansive and cold.

Astronomers use information from surveys across very large areas of the sky to measure the scale of baryon acoustic oscillations imprinted in the universe today. In 2005, astronomers processed the data of more than 45,000 galaxies observed with the Sloan Digital Sky Survey and calculated the baryon acoustic peak to be a little less than 500 million light years.

Recombination

The electrons in the universe join with the protons and nuclei, creating the atoms that comprise all that we are today.

If you put gas into a sealed container and heat the gas to many thousands of degrees, a process called ionization occurs. The electrons will separate from the atomic nuclei they originally orbited, resulting in an electrically charged gas called plasma that consists of the free-flying, negatively charged electrons and the positively charged ions that contain the nuclei. If you then let the gas cool down, the free electrons will join themselves back with the ions – albeit not necessarily the same ones they were orbiting before – and the gas will become electrically neutral once again. We call this process recombination.

Early in cosmic history, the whole universe was like that laboratory container, with electrons and atomic nuclei zipping around in the high-temperature plasma that occupied space. As cosmic expansion continued into the tens of thousands of years, temperatures at last dropped low enough for electrons and ions to begin to attach and form neutral atoms. And even though the particles are all combining for the first time, scientists still call the process by its laboratory name: recombination.

The ambient temperature of space determined which particles recombined first – which was, in turn, determined by how big the universe had grown. After Big Bang nucleosynthesis, one-quarter of the baryons in the universe existed in the form of helium nuclei; they had two protons each, so they needed two electrons to create neutral helium atoms. One electron could recombine with helium nuclei when the universe reached about 30,000 years of age. A second electron could recombine with those single-electron helium ions about 120,000 years after that.

A quarter of a million years after the Big Bang, individual electrons began to combine with individual protons, creating neutral hydrogen atoms. It took a while longer, but by about 380,000 years after the Big Bang, cosmological recombination was just about complete; almost all the matter in the universe was now in the form of atoms.

At this stage of cosmic time, the redshift of the universe (denoted by the letter z) corresponds to the time in years following a straightforward mathematical formula. Using redshift to denote time's passage, the first recombination of helium began to occur at about $z = 6,000$, when the observable universe was one six-thousandth of its current diameter; the second recombination of helium started around $z = 2,000$; and the end of hydrogen recombination happened around $z = 1,100$.

BIOGRAPHY

In the late 1960s, two teams of scientists led by **Jim Peebles** (b. 1935) and **Yakov Zel'dovich** (1914–1987) computed the recombination history of hydrogen in the early universe. They independently deduced that recombination didn't happen in one fell swoop, but rather in a long series of steps taking many thousands of years.

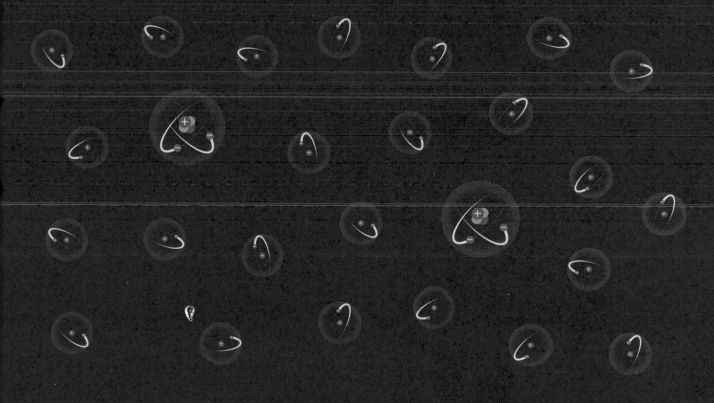

Amid a subatomic sea, alpha particles recombined first with one electron,
and then later with a second electron to create neutral helium atoms.

When recombination finished, hydrogen
atoms outnumbered helium atoms 12 to 1.

200,000 years

Decoupling

Matter and energy part ways and begin to move through space without dragging on each other, allowing light to shine.

Why does it take so much more effort to swim from one end of a pool to the other compared to walking that same distance alongside it? The answer is straightforward: the water drags on your body as you move through it. You're much larger than the molecules of water, fortunately, so eventually you can propel yourself through to the other side. Now, what if you were about the same size as those molecules? Imagine, for example, that you're running from one side of a gymnasium to the other, but the room is filled floor to ceiling with big, glue-covered beach balls. Needless to say, you probably wouldn't get very far.

This is the quandary that light faced in the early history of the universe. Photons that tried to travel through space had to penetrate a dense soup of electrons, atomic nuclei and other particles. In some cases, the photons would collide with a particle, gain or lose energy, and bounce off in another direction; in other cases, the photon would be absorbed completely by an ion or atom – to be re-emitted perhaps, with a different wavelength or direction of motion, but ultimately with no net forward progress.

Once again, the expansion of the universe changed this state of affairs. When light and matter particles are all jammed together and at roughly the same temperature, they drag upon one another and cannot move independently of one another. As the cosmos grew in size, however, the average temperature dropped and the space between particles increased. The universe reached a critical threshold 380,000 years after the Big Bang; the photons and the baryons fell out of thermal balance, and light started to be able to stream past the protons and neutrons – an event astronomers call photon decoupling. Using our gymnasium analogy, the building has been puffed up to such a large size that there is now plenty of space between the beach balls, and you can run across the room without getting stuck.

Coinciding almost exactly with the epoch of photon decoupling is the epoch of recombination, when electrons and nuclei come together to form atoms. Although the two processes differ greatly, they wound up complementing one another; recombination reduced the total number of matter particles that could impede light's travel, and created a large number of new, lower-energy photons that could now move through the universe.

250,000 years

When an electron recombines with an atomic nucleus, it starts off with a large amount of energy. It releases that energy in a series of steps, growing more and more tightly bound to the nucleus as the atom forms. Each step down for the electron releases a photon; and since recombination occurs just as decoupling is also happening, these photons stream into and occupy space, joining the photons that were already there long ago but unable to fly freely until that time. The ratio of matter particles to light particles in the universe since decoupling has been preserved to this day – a little less than one atom per billion photons.

350,000 years

Cosmic Microwave Background

The oldest light in the cosmos, this almost perfectly smooth distribution of energy is the last surviving observable remnant of the early universe.

As photons finished decoupling from matter 380,000 years after the Big Bang, light streamed freely through space. In fact, it fills all of space, which is glowing at an average temperature slightly higher than 3,000 Kelvin (about 3,300°C/5,900°F) That's a bit hotter than the filament inside a typical incandescent light bulb, which glows with an orange-red colour when lit. Had we been around to see it, the universe at that time would have been unbearably bright and hot in every direction.

Yet again, the expansion of the universe changes the conditions within it. The larger it grows, the cooler it gets; at the same time, the wavelength of the background light is stretched according to the cosmological redshift. It took time, but after many millions of years, the temperature dropped to below freezing, and then went even lower, while the visible light in the background stretched into infrared light, and then into microwave radiation. Today, 13.8 billion years after the Big Bang, each photon's wavelength has increased more than one thousand-fold; and nowadays the Sun far outshines those primordial microwaves, which are barely detectable as faint static on old-fashioned, analogue televisions.

Nevertheless, the cosmic microwave background radiation still fills the universe – and it represents the oldest light we humans can detect. This radiation keeps deep space from chilling down to absolute zero; right now, its temperature is just under 3 Kelvin (–270°C/–455°F). And when astronomers discovered the existence of the background, it confirmed that the theory of Big Bang cosmology was correct.

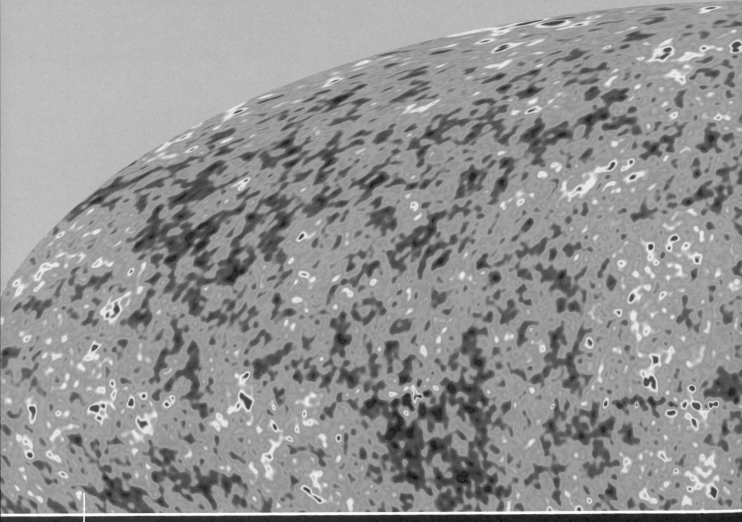

The cosmic microwave background is the boundary between an opaque universe and a transparent one. Before photon decoupling, standing in space would have been like standing in a thick, glowing fog. If the fog suddenly were to clear up, the foggy glow from very distant parts of the universe would take time to reach us; light travels fast, but not infinitely fast. So observing the background is much like looking at a cloud – it has a surface where light cannot penetrate, but that surface isn't a physical barrier. Rather, it's the border around the cloud where incoming light is scattered away and reflected back to our eyes.

BIOGRAPHY

In 1964, American astronomers **Robert Wilson** (b. 1936) and **Arno Penzias** (b. 1933) used a sensitive microwave antenna as a telescope to detect microwave radiation from interstellar objects. To their surprise, they found a pervasive background signal coming from every direction in space. They consulted with their colleagues **Robert Dicke** (1916–1997), **Jim Peebles** (b. 1935), **Peter Roll** (b. 1935) and **David Wilkinson** (1935–2002), and then the two groups announced the discovery and their interpretation of the signal as the cosmic microwave background and direct evidence of the Big Bang.

The Origin of Cosmic Structure

The cosmic background radiation permeates all of space.

When the background was formed just after photon decoupling, all the matter in the universe existed in a blisteringly hot, ionized, gaseous state. Had the gas been uniformly hot and fully mixed, the cosmos may have stayed that way indefinitely, leading to an expanding universe that looked like a cloud for all time.

Instead, the gas was ever so slightly lumpy – like a well-stirred cup of hot cocoa, with just a few flecks of chocolate here and there. By studying the distribution of hot and cold patches in the cosmic microwave background, astronomers have found that minuscule variations of density were embedded in the gas 380,000 years ago. The variations were tiny indeed; imagine a perfectly smooth sheet of ice the size of Paris, with just a few grains of sand scattered on it.

Those humble excesses of matter and energy, it turned out, were more than enough to trigger the grand transformation of the contents of space for the rest of time. When decoupling freed light from the dragging effect of matter, it simultaneously freed matter from the disrupting grip of light. Unlike light, which has no mass, matter creates gravity. Particle for particle, gravity is by far the weakest of the four fundamental forces; so while photons held their sway over the motion of baryons, gravity's influence was essentially unnoticeable. Once decoupled from light, gravity began to dictate the motion of massive particles. Material flowed towards the denser regions and away from the sparser spots. Over billions of years, as the universe continued to expand, matter formed vast gravitational networks across space, organizing matter into filaments, sheets and voids. Today, while the cosmic background takes the form of microwave radiation, space is permeated with a large-scale structure and populated with gas, galaxies, stars, planets, people and so much more.

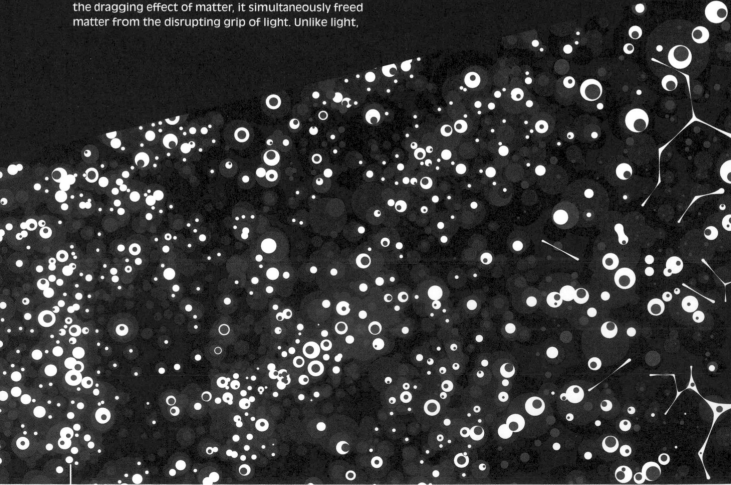

500,000 years

Measurements like those made with the Sloan Digital Sky Survey, depicted below here as an artist's conception, show how galaxies are distributed unevenly through space, in vast webs and filaments surrounded by voids. Using supercomputers to calculate the effects of gravity over billions of years, astronomers have produced simulations of how that large-scale structure of the 'cosmic web' could have resulted from the tiny anisotropies indicated by the unevenness of the cosmic microwave background radiation.

How did the original bumps, or anisotropies, of high and low energy form in the primordial gas? They could have been the result of extremely small quantum fluctuations that arose as the Big Bang occurred and were preserved after the Planck time – and then were vastly expanded and permanently imprinted by inflation. Baryon acoustic oscillations also created ripples in the plasma, which led to regions of higher and lower density when those oscillations were frozen out.

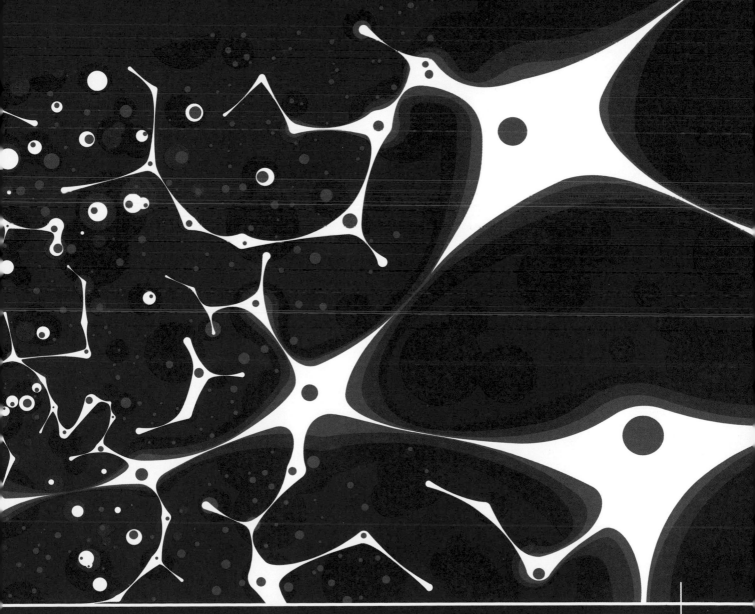

100 million years

4

400 Million Years After the Big Bang

Space is Illuminated as Energy Radiates into the Universe

Light has been unchained from matter, and matter from light.

Four hundred thousand years after the Big Bang, it is finally possible for a photon to travel long distances without being scattered by the vast numbers of electrons that used to fly freely around the universe before recombination.

Meanwhile, the universe continues to expand. A hundred million years later, the universe has increased its diameter by a factor of nearly 50, and the cosmic background radiation is everywhere.

Yet the universe is dark. If at this moment we humans had been there, we would have seen nothing in any direction.

Several effects conspire to cause the darkness. First, as the size of space increases, the temperature of space drops in tandem. At recombination, the universe was on average more than 3,000°C (5,400°F), hotter by far than a blast furnace. Now, its temperature is a frosty −200°C (−328°F). Thus, all the cosmic background is infrared radiation, with wavelengths too long for our eyes to detect.

As for shorter-wavelength radiation, such as visible light, recombination poses a different obstacle. When electrons combine with the products of Big Bang nucleosynthesis – that is, protons, deuterons, helium nuclei and the like – they form neutral atoms that can absorb and scatter higher-energy photons that strike them. It's sort of like standing in a fog bank at night; if you turn on a torch, its light cuts a path into the fog and you can see the beam, but it only travels a short distance before it is seemingly swallowed up by the darkness. The neutral atoms – mostly hydrogen – are spread far apart, but the universe is huge, so every visible photon goes only a small fraction of the distance across space before it hits enough neutral atoms to stop the photon in its tracks.

Happily, help is on the way for all the obstructed light. The decoupling of matter and energy allows a new process to begin: matter, no longer tied to light in its motion, can now move under the influence of gravity alone. The process is slow, but inexorable. Gradually, gas collects into clouds that compress and heat up until they produce bright sources of light – stars! Much of their energy is emitted in the form of powerful ultraviolet radiation; and when those photons hit neutral atoms, they are not absorbed – they split the atoms back into their component electrons and nuclei.

Those first stars in the cosmos soon start to produce another, even more exotic kind of object. Black holes, born in the hearts of the most massive stars, have such strong gravity that not even light can escape from their surface. Matter falling onto those surfaces, however, reaches such high speeds and temperatures that they glow brightly with ultraviolet, X-ray and gamma ray radiation, firing out fierce beams of charged particles that further pierce the haze.

Four hundred million years after the Big Bang, the stars and black holes have done their work to burn away the fog. At this time, if light is produced, it stands a good chance of spreading far into space. Perhaps billions of years later, that light just might strike the surface of a planet and be seen – along with the light of thousands of other stars collected together by gravity into the first galaxies – as the subtlest smudge of light on an eye gazing into the night sky.

The First Stars

Almost immediately after recombination, about 400,000 years after the Big Bang, gravity's influence on matter begins to take hold.

The reverse is also true; a collection of matter actually causes gravity to exist by stretching space inwards towards it, creating the equivalent effect of an ever-pulling force. In the locations throughout space where a tiny bit more matter was present than in its surroundings, more matter collected, first into tiny clumps, and then into larger and larger clouds.

Just about all of the mass that existed in this early cosmic epoch consisted of hydrogen and helium atoms. As these clouds grew ever more massive, attracting more and more sparse gas into them, the clouds themselves started to collapse under their own weight. Layer after layer of the gas fell from the outer parts of the clouds into their cores; and just as a bicycle tyre grows hot as it is inflated, so the temperature of the clouds skyrocketed as the internal pressure grew and grew.

At a critical point, the temperature at the cores of some of these massive, collapsing gas clouds surpassed 10 million degrees and the pressure exceeded 1 billion times that of a car tyre. These conditions existed once before, just a few minutes after the Big Bang – remember what happened then? Nucleosynthesis – protons came together to form deuterons, helium-3 and helium-4 (alpha) particles.

When this process begins in the gaseous cores, energy pours forth from the core. A steady pressure results, which arrests the collapse of the cloud and pushes outwards against the surrounding gas. This cloud is not a closed container, though; imagine trying to inflate balloons that have tiny holes all over their surface. So as long as the energy keeps coming out from the middle, the gas will stay in equilibrium as roughly spherical balls, while heat and light exit the spheres in every direction. The first stars in the universe have been born.

When four hydrogen nuclei (protons) fuse, the resultant single helium-4 nucleus has only 99.3 per cent of the mass of the original protons. The remaining 0.7 per cent of mass is converted into energy, following Einstein's famous formula $E=mc^2$. That rate of efficiency may appear to be tiny; actually, it is far more powerful than any kind of fuel we use on Earth. All of humanity's energy production in the world each year – primarily generated by burning billions of tons of oil, coal and natural gas – could be produced by converting a single hot tub full of seawater into energy.

Hydrostatic equilibrium is the term used to describe objects that have a balance between the gravity that pulls their internal material inwards and the pressure that pushes that material outwards. The material could be gas, liquid, solid or even a mixture of these. In the universe, objects in hydrostatic equilibrium include stars, planets and even large moons and asteroids; they always have a roughly round shape, and are denser in the middle than they are at the surface.

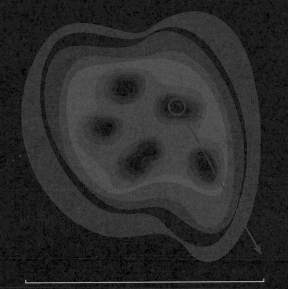

The particular process of nucleosynthesis that occurs in the first stars is called the proton–proton chain, so named because protons interact in a chain of steps to create helium nuclei. Smaller particles are fusing together to form larger, more massive particles; this process is thus also known as nuclear fusion.

1 Astronomical Unit (AU) = 150,000,000 kilometres/ 93,000,000 miles

200,000 AU

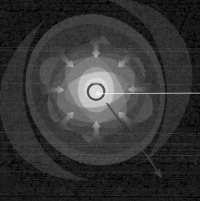

Increasing amounts of infalling matter raise the temperature and pressure at the stellar core

10,000 AU

Gravity causes an accreted disc of material to fall to the centre, where nuclear fusion begins

Nuclear fusion powers jets of matter and energy from the centre perpendicularly outward from the star

500 AU

200 million years

The First Black Holes

The first stars in the universe are brilliant and massive – up to dozens or even hundreds of times the mass of our star, the Sun.

The most massive of these stars shine millions of times more brightly than the Sun ever will. That great luminosity, however, exacts a cost: the nucleosynthesis at their hearts, which produces that brightness, works so rapidly that these stars consume their raw material – the protons that fuse into helium – in a fraction of the time.

Within a few million years at most, low on hydrogen and choked with too much nucleosynthetic by-product, the conditions in such a stellar core can no longer sustain enough nuclear fusion to maintain the delicate balance between its outward push and the steady inward press of gravity. At the critical moment, the core collapses, producing a new kind of object in the universe – one whose gravity is so strong that, once inside its boundary, not even light can escape – a black hole.

From the outside, black holes act like any other dense body in the universe, exerting gravity on its surroundings. Contrary to popular misconception, they do not act like cosmic vacuum cleaners; in other words, they do not 'suck' matter any more vigorously than any other objects of the same mass, such as stars or planets. Rather, they are so compact that outside matter can fall closer to the centre of a black hole without going inside than to any other kind of object. As it arrives at the black hole's 'surface' – a threshold called the event horizon – any infalling material will reach the speed of light, which is the speed limit of anything moving through the universe. Once something falls into the black hole, it's not coming out again.

BIOGRAPHY

In 1783, the English scientist and clergyman **John Michell** (1724–1793) was the first person to calculate the parameters of an object with such strong gravity that light could not escape its surface. He suggested that such 'dark stars' could be numerous in the universe. His conclusions were echoed in 1795 by the French astronomer and mathematician **Pierre-Simon Laplace** (1749–1827). Scientific research on black holes stopped there, however, and did not resume for more than a century. The term 'black hole' was likely first used in scientific conversation in the 1960s by American physicist **John Archibald Wheeler** (1911–2008), and it made its first appearance in print in an article written by American science journalist Ann Ewing in 1964.

The English mathematical physicist **Roger Penrose** (b. 1931) is known for some of the most creative thinking ever applied to the way abstract mathematics expresses itself in the physical universe. In 1965, while a reader at Birkbeck College, University of London, Penrose published an article explaining that one consequence of Albert Einstein's general theory of relativity would be the formation of a 'singularity' in space by the gravitational collapse of a massive stellar core. His dizzying and beautiful mathematical analysis laid the theoretical foundation of black holes, and in 2020 earned him the Nobel Prize in Physics.

Giant first-generation stars in the universe evolve into supergiants, which then explode as supernovae, leaving behind a totally collapsed core: a black hole so dense that not even light can escape its gravitational pull.

Stage 1 – Giant star

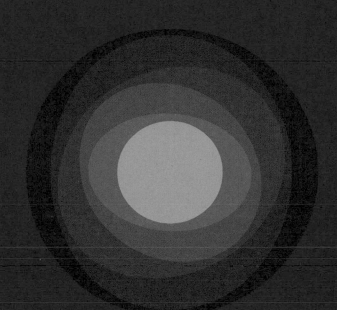

Stage 2 – Supergiant

Stage 3 –Type II supernova

Stage 4 – Black hole

300 million years

Black Holes in the Cosmos

Black holes, by cosmic standards, are physically puny – but they pack quite the gravitational punch.

Singularity

Event horizon

The outer boundary of a black hole – the 'surface', so to speak, from where you cannot escape once you land – is called the event horizon. If our planet Earth were suddenly to become a black hole, the event horizon would be about the size of a golf ball.

When an actual golf ball is hit into the air, it lands on the green with a downward velocity of perhaps 30–40 metres per second (67–90 miles per hour). If the ball were hit off the edge of a canyon, however, it would be going much faster when it reached the bottom. What would happen to a golf ball, then, that fell all the way down to the black-hole Earth? It would speed up, faster and faster, and land on the event horizon at the speed of light, which is an astonishing 299,792,458 metres per second (670 million miles per hour). The golf ball would be converted from mass to energy, producing an explosion 100 times more powerful than the atomic bomb detonated in 1945 over Hiroshima, Japan. But all of that energy would be absorbed into the black hole, invisible to all who are watching.

Realistically, though, the likelihood of the golf ball surviving the trip would be quite slim. If there were any amount of gaseous material surrounding the black hole, like the atmosphere that covers Earth's surface, the golf ball would almost certainly get superheated by the friction it experiences and be vaporized into atoms long before it reached the event horizon. The leftover heat would make the gas hotter, ultimately leading it to glow with radiation; and this leads to the cosmic irony that the environment around the darkest objects in the universe can often shine brighter than the hottest stars.

When a bath full of water is draining, it cannot all fit down the pipe at once. Instead, the water flows slowly towards the drain, creating a swirling disc that gradually lets the water into the plumbing. Much the same thing occurs as matter falls into a black hole; the final destination is so small that matter piles up in a swirling accretion disc, waiting for its turn to reach the black hole's surface. The disc gets so hot – reaching temperatures of millions or even billions of degrees – that most of the matter never gets there; instead, the super-energized material jets outwards, perpendicular to the accretion disc, blasting everything in its path with powerful radiation.

This first direct image of the closest surroundings of a black hole was obtained in 2019. More than 6 billion times the mass of the Sun and 53 million light years away from Earth, the black hole is hidden behind, and is about one-third the diameter of, the dark spot surrounded by the orange glow. An international team of hundreds of scientists used the Event Horizon Telescope – a network of observatories around the globe – to gather radio wave emissions from the black hole and its surroundings. All the separate, partial pictures were then scientifically synthesized by the American computer scientist Katie Bouman (b. 1989) into a single, final image.

What does a black hole look like inside the event horizon? Based on theoretical models, there should be a singularity at the centre – a point of zero volume and infinite density, acting much like a pinprick in the structure of space. In all the volume between the singularity and the event horizon, there could be more detailed structure – but from our vantage point outside the event horizon, we cannot see or detect any of it.

The region around a supermassive black hole, viewed with the Event Horizon Telescope

Active galactic nucleus with an accretion disc and jets

Model edge-on view of the supermassive black hole region

400 million years

The First Galaxies

As matter continues to aggregate under gravity's influence, an invisible yet plentiful constituent of the universe produces a powerful impact.

In the several hundred million years after matter decoupled from energy throughout the cosmos, the protons, neutrons and electrons that make up all the familiar matter on Earth have interacted with gravity, electromagnetism and the nuclear forces to make stars and black holes. Scientists call this material 'baryonic matter'. Around this time, matter of another kind starts to make its presence felt as well.

We still don't know what kinds of particles this other type of matter is made of. What we do know, however, is staggering. There is more than 5 kilograms of this other matter for every kilogram of baryonic matter in the universe, yet it is completely dark. It's not dark like a black hole, whose light is trapped within its event horizon; instead, this cosmological dark matter appears only to interact with gravity, so it simply emits no light or radiation of any kind no matter where it is.

So where is it? Dark matter started out with the same, nearly uniform distribution throughout the universe that baryonic matter had. Once freed from being dragged around by light, dark matter also started to gather into structures; without the influence of electromagnetic and nuclear forces, though, the process was much slower and gentler. So, in the course of hundreds of millions of years, dark matter formed sparse agglomerations thousands of light years across, while the baryonic matter that mixed in with the dark matter came together to form gas clouds, some of which would make stars, and some of which would further make black holes.

Half a billion years after the Big Bang, concentrations of luminous baryonic matter have fallen together into the centres of these structures, cocooned in haloes of dark matter. Over time, they will further evolve in shape and size as the forces of nature work on them and their components, creating all kinds of forms ranging from irregular streaks to dwarf ellipsoids and grand spiral discs. The first galaxies have been born.

The first galaxies in the universe are so faint and far away that they look like fuzzy blobs (inset images on the facing page) even with our most powerful telescopes. New observatories, including the James Webb Space Telescope, will use advanced infrared technologies that will allow us to see these galaxies in a new light.

On average, a volume of space the size of our planet Earth would contain about one billionth of a gram of dark matter. At cosmological sizes and distances, however, that tiny density is more than enough to pull galaxies together and shape the large-scale structure of the universe.

BIOGRAPHY

Dark matter's existence was first noticed in the 1930s by Swiss-American astronomer **Fritz Zwicky** (1898–1974) when he observed that galaxies in a large cluster were moving faster than they should have been. Four decades later, American astronomer **Vera Rubin** (1928–2016) showed that the motion of the outer regions of galaxies was also much faster than expected. Rubin showed that these motions could be explained by the presence of large amounts of unseen matter.

500 million years

Reionization

Powerful electromagnetic radiation has permeated the universe to allow light to shine all the way across it.

Atoms – the result of negatively charged electrons and positively charged nuclei combining to create a single composite particle – have the remarkable property of being able to absorb light that strikes them, and later emit that light again in different colours. The exact colours and wavelengths of light that any given kind of atom can absorb or emit are determined by its quantum mechanical properties; if there are enough atoms, though, in the line of sight between a light source and your eye, there will be broad patches of colour in which you won't be able to see the light at all.

This is exactly what happens in the universe after recombination. Light is untethered from matter, thanks to the decoupling that happens at just about the same time, 400,000 years after the Big Bang; but even though the density of atoms in space is low, there are so many of them that it's only a matter of time before much of the light from any source is extinguished. Sure, it can travel thousands or even millions of light years; but by the time the first stars have formed and are shining, the observable part of the universe is already 100 million years across and growing fast.

The way to make space transparent to visible light once more is to separate electrons from their nuclei again. When the electrically charged nuclei, called ions, are stripped of their electrons, they are also stripped of their ability to absorb light; and now that the temperature and density of space is much lower than it was before recombination, light will hardly be scattered by free-floating electrons as they travel. So the universe, once ionized because of its compactness and high temperature and rendered neutral at recombination, now must be reionized.

Reionization takes much longer than recombination, because there are few sources of ionizing radiation at that period in cosmic history. Eventually, though, it happens. For hundreds of millions of years, massive stars radiate large amounts of ultraviolet radiation into space, striking neutral atoms and breaking them apart. In parallel, black holes convert the energy of infalling matter into ultraviolet, X-ray and gamma ray radiation, and send magnetically coiled jets of matter and light in powerful beams far into the cosmos. About 550 million years after the Big Bang, more than 99.9 per cent of the volume of space has been ionized. The remaining neutral gas has gathered into clouds the size of galaxies or smaller – not large enough nor numerous enough to block the view. It is now possible to see the marvels of the cosmos, all the way across the observable universe.

Does cosmological dark matter, which far outweighs the charged ions and electrons, affect the transparency of space? No; although dark matter can change the direction of light by curving the space light travels through, it doesn't block light from travelling past it – indeed, it doesn't interact with light at all.

600 million years

800,000,000 years

5

4 Billion Years After the Big Bang

Galaxies Are Born

If a black hole is gravity's temple, then a galaxy is gravity's playground.

Under gravitational influence, billions of suns' worth of matter stream towards a common location in space. Left unchecked, they would fall together to form a black hole; when other physical processes arrest that development, the result is a beautiful and dynamic collection of material, surrounded and stabilized by cosmological dark matter.

Galaxies form gradually, but the process can be either smooth or lumpy. Supercomputer simulations have shown that if the bulk of a galaxy's material streams gently together – perhaps a few hundred or a few thousand solar masses per year, over hundreds of millions of years – the physical property known as the conservation of angular momentum will lead to the formation an orderly, swirling disc surrounding a central bulge. The rotation of the matter in the disc keeps it from falling any further into the galaxy bulge, in much the same way that water in a child's beach bucket won't pour out if it's swung around with enough speed. Within the disc, the orbiting matter interacts further via friction and gravity, producing the beautiful carousel-like variations we see in disc galaxies as, over many thousands of millennia, mostly-spherical bulges become elongated bars and then turn back into bulges; spiral arms form, dissipate, wind and unwind; and billions of stars are born, live, reproduce and die.

When a galaxy accretes a large chunk of matter all at once – a smaller proto-galaxy, perhaps, gravitationally distinct but not yet fully formed – the result can be dramatic. The sudden addition of so much new material disrupts the orderly disc, as gravitational tides tear through the system to create warps, loops and tails. The galaxy puffs up into a rugby ball-shaped collection of stars moving in random directions – a benignly buzzing cosmic bee swarm trying to regain its bearings. Left undisturbed for long enough, this new elliptical galaxy will settle back into a disc; if it keeps being hit by large clumps, however, it could retain its ellipsoidal structure indefinitely.

And what of those quasar-spawning supermassive black holes that are so ubiquitous during this era of cosmic history? Every large galaxy in the universe that we've had the technology to measure contains a supermassive black hole at its centre, although only a few per cent of them are accreting matter actively enough to become gravitationally powered super-engines known as quasars. It's thus tempting to imagine that black holes act as gravitational seeds around which matter gathers and galaxies form. Yet studies also show that galaxies can form just fine without a central black hole, and no supermassive black hole has yet been found that isn't located in a home galaxy. So is the galaxy the chicken or the egg, and is the black hole the egg or the yolk? That's one of the outstanding mysteries of astronomy today.

Galaxy Formation

Structure in the universe forms hierarchically.

In other words, little things form inside big things while those big things are still forming themselves, while still larger objects are slowly forming that contain the big objects. As gravity pulls the constituent parts of a galaxy towards a central spot, stars have already formed within subgalactic clumps of baryonic and dark matter.

Those subgalactic blobs are usually formidable enough structures to be called dwarf galaxies themselves. They often already contain millions of stars' worth of gas and dust before they start linking together. Those first larger galaxies look like chains, bugs or even cosmic tadpoles. Like all galaxies, though, they're bound together almost completely by gravity and nothing else; so rather than being solid objects, the galaxies' different parts are still moving relative to one another, continuously changing positions and shapes – it's just happening too slowly for us to notice.

Let's say enough matter has collected to create a big, beautiful galaxy. How long does it take to put that galaxy together? Although the collecting material could fall towards a common centre at speeds exceeding 1 million kilometres per hour, it needs to fall 1 million trillion kilometres or more to reach the middle of the maelstrom. Forming a galaxy thus takes billions of years; and during that time the smaller, infalling pieces of the galaxy will themselves evolve, as gas clouds and stars and star clusters form and interact within them.

The oldest galaxies observed by astronomers formed about 1 billion years after the Big Bang. In order for us to see them in their youthful stages, they have to be so far away that light coming from them has to have travelled almost 13 billion years. This phenomenon, called 'look-back time', works for any distant object in the universe and gives us a direct view of cosmic history unfolding before our eyes – although we are left to wonder: what's going on over there right now? We won't know until the light these distant galaxies emit now reaches us, far in the future.

The tadpole-like dwarf galaxy known as Kiso 5639 may be an example in the local universe of what many galaxies looked like in the early history of galaxy formation. The limits of astronomical technology – for example, this deep field image (below left) taken with the Hubble Space Telescope – make Kiso 5369's distant counterparts look like odd organisms.

Kiso 5639

The Hubble Ultra-Deep Field covered a patch of sky only one-hundredth the size of the full Moon. More than 600 hours of viewing time with the Hubble Space Telescope revealed the presence of some 10,000 galaxies in that tiny area, including many caught in the act of formation.

1,200,000,000 years

The Age of Quasars

As galaxies formed, the black holes within them began to grow in both number and mass.

The most massive black holes fell to the centres of their galaxies; and in those galaxies where gas and stars could also fall into the centre, they piled up around the event horizon of their central black hole like floodwater entering a storm drain. Some of the matter fell into the black hole, making it larger and more massive; the remaining material that gathered, supercharged by powerful electromagnetic fields and superheated to billions of degrees, blasted out of that central region before falling all the way in. Matter was carried away at close to the speed of light in vast, fountain-like jets, and the entire region shone with enormous energy.

If the rate of the flow of matter into a galaxy's central black hole exceeds about a solar mass per year (about 40 Earths per hour), the energy released is so great that it can easily outshine all the stars in its galaxy combined – not only with visible light, but also with all kinds of electromagnetic radiation, from radio waves to gamma rays. The first such black hole-powered objects were discovered in the late 1950s using radio telescopes – yet they looked like single stars through an optical telescope. They were dubbed 'quasi-stellar radio sources' or quasars for short. It took decades' more study before astronomers confirmed that quasars have black holes for hearts – and supermassive ones at that, ranging from millions to billions of times the mass of our entire solar system.

There are relatively few quasars in the universe today compared with the distant past, implying that the conditions for growing supermassive black holes rapidly were much better back then than they are now. Thanks to their brilliant luminosity and compact size, quasars act like beacons, marking distances far across the cosmos. When quasars' light shines through transparent clouds of intergalactic gas, astronomers can observe the markers of those clouds in the quasar's spectrum, revealing baryonic matter that we cannot otherwise see.

Astronomical surveys have shown quasars were most numerous during the time around 2–3 billion years after the Big Bang. During that period in cosmic history, galaxies were forming at a rapid rate, and a large amount of free-floating hydrogen gas was available to fall into the centres of galaxies, which probably helped fuel quasar activity.

Quasars such as 3C 273 are gravitational engines powered by their supermassive central black holes. A large percentage of the mass that falls into a quasar is converted to energy and emitted into space without disappearing past the black hole's event horizon, making them much more powerful than even the nuclear fusion engines inside the hearts of stars.

Quasars are one kind of active galactic nuclei (AGN) – so named not because of nucleosynthesis or nuclear fusion, but because they contain supermassive black holes located at the centres (nuclei) of galaxies. Quasars are some of the most luminous AGN in the universe, and are known for producing vast amounts of radio waves when their prodigious energy interacts with the gas that surrounds them within their host galaxies. Hercules A, shown here as viewed with a radio telescope, is a dramatic example of this phenomenon, powering jets of emission that blast far beyond the visible limits of its host galaxy.

BIOGRAPHY

Dutch-American astronomer **Maarten Schmidt** (b. 1929) was the first person to measure the distance between Earth and the quasar 3C 273 – more than 2 billion light years. That meant that quasars could not be single stars. Decades later, astronomers confirmed that 3C 273 was a supermassive black hole system so bright that it drowned out the light of its host galaxy.

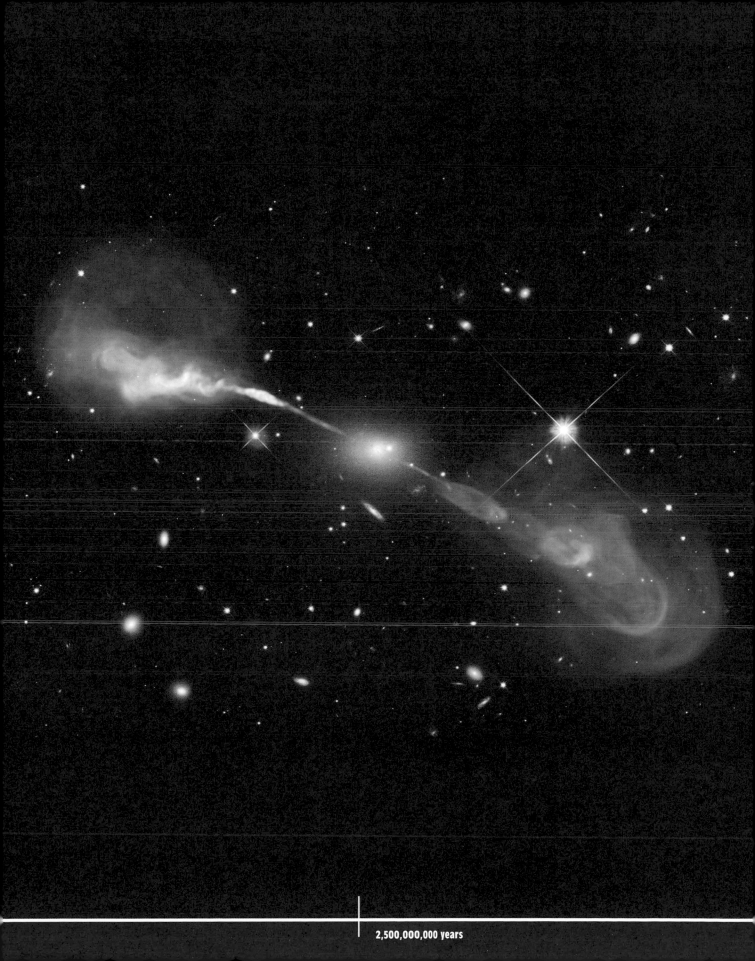

2,500,000,000 years

Peak Star Formation

As gas falling into galaxies fuels powerful quasar activity in galaxy cores, it also provides fuel for star formation.

Left alone, an interstellar gas cloud rarely begins to collapse by itself; on the other hand, if it's stimulated by another cloud ramming into it, or perhaps buffeted by a quasar's shock wave, that infusion of energy can set into motion the chain reaction of starbirth, as one new star's radiation and charged-particle winds induce another star to form, and then another.

Around 3 billion years after the Big Bang, the rate of new stars being created from gas is at its peak in the universe. Galaxies evolve as well, as gravity and rotation work together to organize their stars into two basic configurations. If the galaxy's matter is rotating rapidly, its stars will settle into a disc, with an occasional arm-like pattern of spiral waves running around its centre. If rotation is lacking, the stars will fall further towards the centre, creating a spheroidal or rugby ball-shaped bulge of stars that zip around and through the galaxy core in nearly random directions like hornets around a hive. Combinations occur, too, resulting in lens-shaped lenticular galaxies and irregular systems either too small to form well-defined structures or still unsettled from recent interactions or collisions.

Over billions of years, the majority of galaxies have congregated into communities, from neighbourhood-sized groups of galaxies to giant clusters of galaxies. The largest clusters can have dense concentrations and suburban outskirts, much like human metropolitan areas, mixed into vast volumes containing hundreds of thousands of members.

Disc-shaped galaxies often have spiral patterns in them that are not solid 'arms', although they look that way; they are denser regions of gas, dust and stars that form in much the same way that swirls form in a draining bathtub – except these galactic swirls last for hundreds of millions of years. Most of the star formation in galaxies today occurs in disc galaxies.

Bulge-shaped galaxies contain relatively fewer young stars than disc-shaped ones, and maintain their shape not by spinning but by more random stellar motions that 'puff' matter away from their cores and into ellipsoidal balls. The most powerful quasars are generally located at the centres of these galaxies.

The majority of galaxies in the universe are small compared to our own Milky Way. They can take on disc shapes or bulge shapes, or they can be irregularly configured, almost as if they haven't had enough time yet to organize themselves into grand designs.

BIOGRAPHY

American astronomer **Edwin Hubble** (1889–1953) was the first to confirm that galaxies like the Milky Way were scattered far and wide throughout an expanding universe. In his book *Realm of the Nebulae* (1936) Hubble described the many shapes of galaxies as fitting into a scheme with two main branches, the spirals and the ellipticals. This 'tuning fork' of galaxy morphology turned out not to be a formation sequence of galaxies, but an organizational method that has helped astronomers understand the way stars are born and distributed in galaxies.

3,500,000,000 years

The Milky Way

Four billion years after the Big Bang, our Milky Way has begun its life.

Today, our galaxy contains hundreds of billions of stars, the gaseous raw material to make billions more stars, and enough heavy-element dust to make trillions upon trillions of planets. The vast majority of the Milky Way's stars stretch out into a spiralling, wavy disc about 100,000 light years across. At the centre is an olive-shaped, elongated bulge that astronomers call a galactic bar – like a plum stuck in middle of a thin-crust pizza – containing billions more stars, all buzzing around an ominous object more than one trillion times the mass of Earth: the supermassive black hole called Sagittarius A*.

The rest of the stars in our galaxy fill out a sparse, spherical halo somewhat larger in diameter than the disc itself, and this halo is peppered with more than a hundred globe-shaped clusters of stars, orbiting the galactic centre like a cloud of satellites. All of this grand structure – a veritable island universe unto itself – lies within a cocoon of dark matter that further extends more than 100,000 light years in every direction.

Our solar system, and its third planet on which we reside, orbits around the galactic centre in a huge, undulating circle, taking about 250 million years to make one full round trip. We're situated not in the middle of the galaxy, but about halfway out from the centre – thankfully far from the central black hole, in a relatively quiet spur of one of the large spiral arms. Left largely undisturbed, we have been able to sustain a steady galactic local environment for billions of years, which has in turn allowed Earth's planetary ecosystem to develop slowly but surely into the comfortable home where we live today.

Ironically, although we can observe galaxies and quasars billions of light years away, large parts of our own galaxy remain hidden from our view, as if we were a minnow in a murky pond. The cloudy material in the Milky Way's disc screens our line of sight in numerous directions just a fraction of the way across the disc. To help pierce the veil, astronomers use infrared technology – that is, heat vision – in telescopes and detectors to gaze beyond what our eyes alone can see.

With a precise measuring technology called interferometry, astronomers using the GAIA (Global Astrometric Interferometer for Astrophysics) space telescope have mapped the precise positions, colours and motions of some 2 billion stars. This is the most detailed picture of our galaxy ever obtained, enabling increasingly accurate models (like the artist's concept shown here) of its full size and shape. The data show that the Milky Way continues to grow and evolve as clumps of matter continually fall into and move around our galactic disc and halo. GAIA maps have been used to map the path of our solar system, and also to predict what our galaxy might look like up to 400,000 years from now.

Earth's location (circled on the Milky Way map on the facing page) is about halfway between the centre and the edge of the spiralling stellar disc of our Milky Way galaxy.

BIOGRAPHY

The Dutch astronomer **Jacobus Cornelius Kapteyn** 1851–1922) used the first large-scale maps of stars in the night sky to deduce that the Milky Way galaxy was disc-shaped. The American astronomer **Harlow Shapley** (1885–1972) used the distances and motions of the globular clusters in the Milky Way's stellar halo to calculate the size of the Milky Way, and to locate the off-centre position of the Sun.

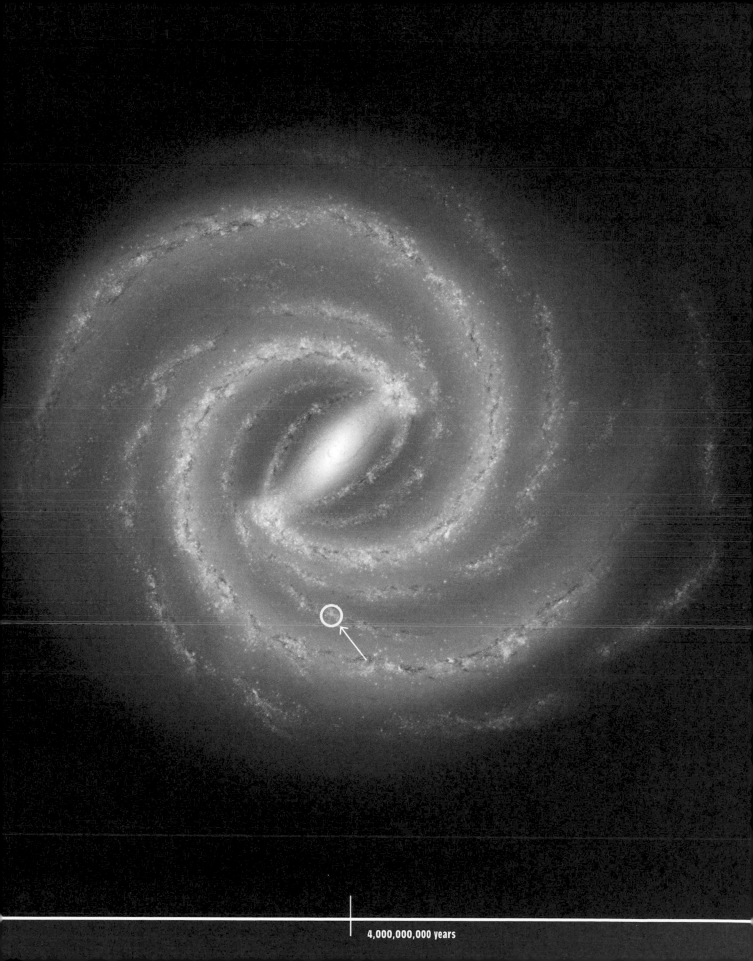

4,000,000,000 years

Galaxy Assembly

Galaxies are to the universe, in a sense, what cells are to the human body.

We humans can only see a small part of the universe in detail; so, just as a medical researcher might study cellular division and cell death to decipher the human ageing process, astronomers examine the formation – and transformation – of galaxies to deduce what might happen to the universe with the passage of cosmic time.

To humans, though, galaxies age and evolve on a timescale so long – millions and billions of years – that no person in a single lifetime could ever watch a galaxy change its shape and structure. It's akin to a mayfly studying the ageing process of a tortoise. Undaunted, astronomers use scientific strategies to peer into cosmic history. One solution is to look at as many different galaxies as possible, and another is to use computers to watch galaxies evolve on fast-forward.

Imagine if you only had one day to figure out how human beings grow old. What could you do? You might take a snapshot of all the people at London King's Cross, and then categorize the different people in the picture – how large or small they are, what they're wearing, whether they're walking or being wheeled in buggies, and so forth. Then you could make educated guesses about how people start off as babies, grow to become children and then adults, and how they look depending on how they spend their time. Your colleagues could then take photos of other gatherings of people at other places: at a school, or a stadium or just down the street from your home. Eventually, a unified picture would emerge of the human ageing process, even though it might not be possible to watch a single person live an entire lifetime in a day.

What if we could speed up time, at least in a simulated universe? Some basic laws of physics are understood well enough that we can use them to predict what might happen far into the future, if only we could calculate all the results fast enough. Scientists study the universe in this way using computers, setting up conditions in the early universe and running time forwards at superspeed to see if they yield the galaxy populations we observe today. It's almost like a cosmic video game – and, indeed, many of the most popular science-fiction movies and computer-gaming platforms today derive their realistic motion and dynamics from the computational engines that were first developed for astrophysics research.

BIOGRAPHY

The German-born British astronomers **William Herschel** (1738–1822) and **Caroline Herschel** (1750–1848) started their careers as professional musicians. After the brother-sister team turned to science full time, they made a long list of important astronomical discoveries, and assembled the first comprehensive list of more than 2,500 non-stellar objects in the universe. William's son **John Herschel** (1792–1871) further extended the catalogue with observations from Earth's southern hemisphere. Their work eventually became the core of the New General Catalogue (NGC) of nebulae, clusters of stars and galaxies, which is still one of the most widely used astronomical catalogues today.

Two small galaxies approaching one another at just the right speed can merge into a larger single system over a billion or more years.

BIOGRAPHY

In 1941, well before computers were first built, the Swedish astronomer **Erik Holmberg** (1908–2000) published a remarkable experiment on the interaction of galaxies using light bulbs on a table and calculations of gravitational forces by hand. Today, astronomers use supercomputers to see in detail the effects that Holmberg correctly simulated eight decades ago.

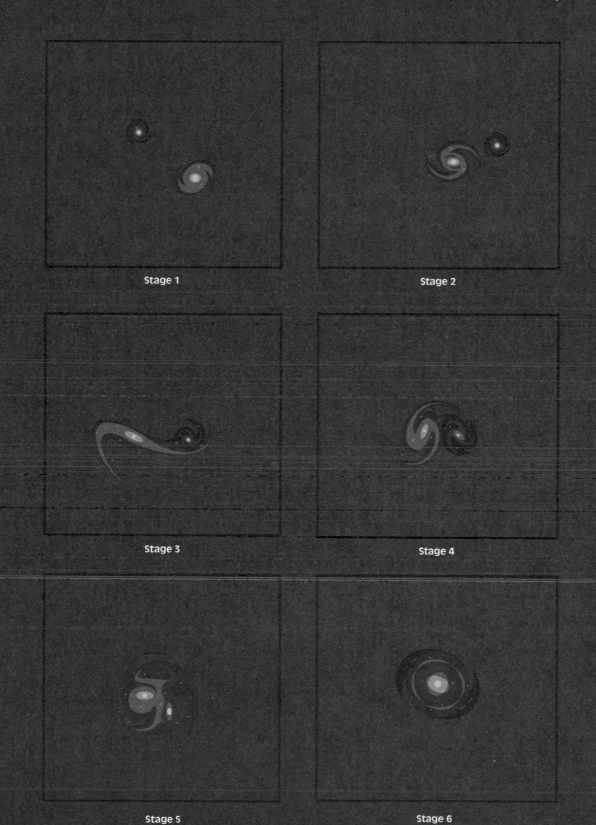

Stage 1

Stage 2

Stage 3

Stage 4

Stage 5

Stage 6

4,750,000,000 years

5,000,000,000 years after Big Bang

6

4.6 Billion Years Ago

Birth of the Sun and the Solar System

More than 5 billion years have passed since our Milky Way was born.

As the halo of matter that cocoons and defines our galaxy grows larger and larger, with the addition of hundreds of smaller halos, the dark matter portion provides an island of gravitational stability while the baryonic matter portion forms vast clouds and streams of gas. And inside those clouds are born billions upon billions of stars.

The original stars that formed in the universe were huge compared to these next generations of stars. Although a few per cent did grow as big as those first-generation giants, later-born stars on average were only a fraction as massive. Why? Those first stars seeded the interstellar environment with the ingredients necessary to make it possible for their smaller descendants to form: carbon, nitrogen, oxygen, iron and more – the elements of the periodic table.

At first glance, the formation of other elements – a new kind of nucleosynthesis, much more complex than that which happened in the first minutes after the Big Bang – seems like a tiny effect. Even after billions of years, hydrogen and helium – the two original types of atoms in the cosmos – still comprise more than 99 per cent of all atomic nuclei; and even though each type of newer nucleus is heavier by far, the two lightest elements still outweigh all the other

elements put together nearly 50 to one. The tiny fraction of heavy elements, though, makes all the difference between the stars born in the Milky Way galaxy compared to those born before the reionization of the universe.

The population of stars now is amazing – diverse and numerous, each one sharing key properties with its neighbours, yet also unique in crucial ways. Stars live now in teeming clusters, or in pairs or triples, or in splendid isolation. They glow in all the colours of the rainbow, as well as wavelengths of light invisible to human eyes. And, depending on their mass and composition and the environment where they were born, they could live on and on – or stop shining in a spectacular final cataclysm, sharing their matter and energy with the rest of the galaxy to give birth to new stars and begin the stellar life cycle anew.

As the Milky Way's stellar metropolis grows and matures, two important milestones mark time's march towards our present moment. The first is time itself, or more accurately, our keeping of it, as the heavy elements in the universe provide a new way to measure the passage of time. The second is, a little more than 4.5 billion years ago, the birth of our own star, the Sun.

The Sun may have formed in a brilliant open cluster of stars like NGC 3603, observed here with the Hubble Space Telescope, surrounded by billowing clouds of gas and dust that served as the raw material for stellar birth.

Stars Populate the Universe

The first stars in the first galaxies helped to reionize the universe about half a billion years after the Big Bang.

Thanks to the nuclear fusion inside them, turning hydrogen into helium while releasing tremendous amounts of heat and radiation, the next generations of stars shone far across the universe, only occasionally impeded by clouds of gas and dust.

Star-forming regions such as CTB 102, about 14,000 light years from Earth, are often obscured from view by thick interstellar dust. Observations with radio telescopes show dozens of young stars enveloped within enough gas to make 100,000 Suns.

The majority of the Milky Way's stars were formed in the few billion years of its life. The stellar birth rate gradually fell as the amount of hydrogen gas available to form stars dropped, and the physical conditions of our galaxy changed and stabilized over time. Today, there are an estimated 300 billion stars in our galaxy, and on average a few more are born every year.

Optimal conditions for star formation often occur in the same spots. Huge amounts of gas congregate inside galaxies in clouds; within these clouds, denser clumps form, and within those clumps even denser cores collapse and form new stars. Our Sun probably formed in a loose association of stars that separated far across the galaxy after they were born. Tighter groupings form open clusters of stars, which stay together for hundreds of millions of years. When many thousands of stars form together in a single cloud, they can stay tightly packed as a globular cluster. The Milky Way has about 100 known globular clusters moving in and around it, some of which contain stars that are older than the Milky Way galaxy itself.

Light from any source follows an inverse square law as it travels; its flux (that is, its brightness as seen from afar) goes down by a factor of four every time the distance is doubled. Although the drop is rapid, it never reaches zero, and is gradual enough that bright sources of light such as quasars and galaxies can be detected almost out to the edge of the observable universe.

The Pleiades (top of facing page) is an open cluster that contains the famous 'Seven Sisters', a configuration of young stars that for millennia have been the inspiration for myth and legend. 47 Tucanae (bottom of facing page), a globular cluster visible in the southern hemisphere, is almost like a tiny galaxy unto itself.

A large fraction of the stars in the galaxy are in binary or multiple star systems. Binary and multiple systems can be visual, where each star can be clearly seen, or spectroscopic, where the additional stars are too close to be distinguished in a picture but can be detected by their motion over time.

BIOGRAPHY

English-born American astrophysicist **Cecilia Payne-Gaposchkin** (1900–1979) was the first recipient of a PhD in astronomy at Harvard. In her doctoral thesis in 1925, she correctly deduced that stars were made almost entirely of hydrogen and helium – a major departure from the thinking of the time that stars had Earth-like composition. Her research laid the groundwork for the study of high-luminosity stars and variable stars. She later became the first woman full professor on Harvard's faculty, as well as Harvard's first woman department chair.

5,500,000,000 years

Stellar Brightness and Colour

Far from being just bright dots in a dark sky, stars display a variety of luminosity (how much energy they produce) and colour.

We humans perceive colours as combinations of various shades and hues, and have a hard time describing just how 'blue' or 'red' objects appear. About two centuries ago, astronomers began to use physics and mathematics to describe the colours of stars, comparing the ratios of their light in specific wavelength bands. Standardizing these measurements allowed us to study stars scientifically and classify the vast panoply of different kinds of stars in the universe.

Examining colours in even greater detail can be more time-consuming, but can also yield great scientific rewards. If light from an object is refracted through a prism or reflected off a diffraction grating, it is spread out into its component colours, creating a spectrum. (A rainbow is the spectrum of the Sun, refracted through a temporary prism created by water droplets in the atmosphere.) In 1802, English physicist and chemist William Hyde Wollaston (1766–1828) obtained the first spectrum of the Sun, and noticed dark regions between the bright swathes of colour; in 1814, German physicist Joseph von Fraunhofer (1787–1826) used improved equipment to distinguish the dark regions as a myriad of narrow lines. Decades later, astronomers figured out that these lines were caused by the interactions between light and the atoms in the Sun.

The colours of stars have a direct connection to their spectral classifications and their temperatures. The hottest stars, with surface temperatures of 30,000°C (54,000°F) and greater, glow violet-blue and are called O stars. B stars are about 20,000°C (36,000°F) and A stars are about 10,000°C (18,000°F); they glow blue and white. F, G, K and M stars are progressively cooler, ranging down to 3,000°C (5,400°F), and glow yellow, orange and red.

In 1911 and 1913 respectively, astrophysicists Ejnar Hertzsprung (1873–1967) of Denmark and American Henry Norris Russell (1877–1957) published plots of stars that visualized the way they are distributed in colour and luminosity. Today, the Hertzsprung–Russell diagram is one of the most useful tools scientists possess to study the properties of stars.

At least nine out of every ten stars in the Milky Way are main-sequence stars, so named because their colour and luminosity land on that curved strip (the main sequence) of the Hertzsprung–Russell diagram. Two other major regions of the diagram are occupied respectively by red giants and white dwarfs – objects whose names aptly describe their appearance and size.

BIOGRAPHY

Three American astrophysicists pioneered the classification of stars based upon their spectra. **Williamina Fleming** (1857–1911), who was born in Scotland, established around 1890 a simple system based on the strength of spectral lines produced by hydrogen. In 1897, **Antonia Maury** (1866–1952) published a catalogue of hundreds of stars based on thousands of stellar photographs, and proposed a system of classifying stars with spectra that emphasized both their brightness and colour. A few years later, **Annie Jump Cannon** (1863–1941) synthesized the work into the Harvard Classification Scheme, the basis of the 'OBAFGKM' sequence that astronomers use today.

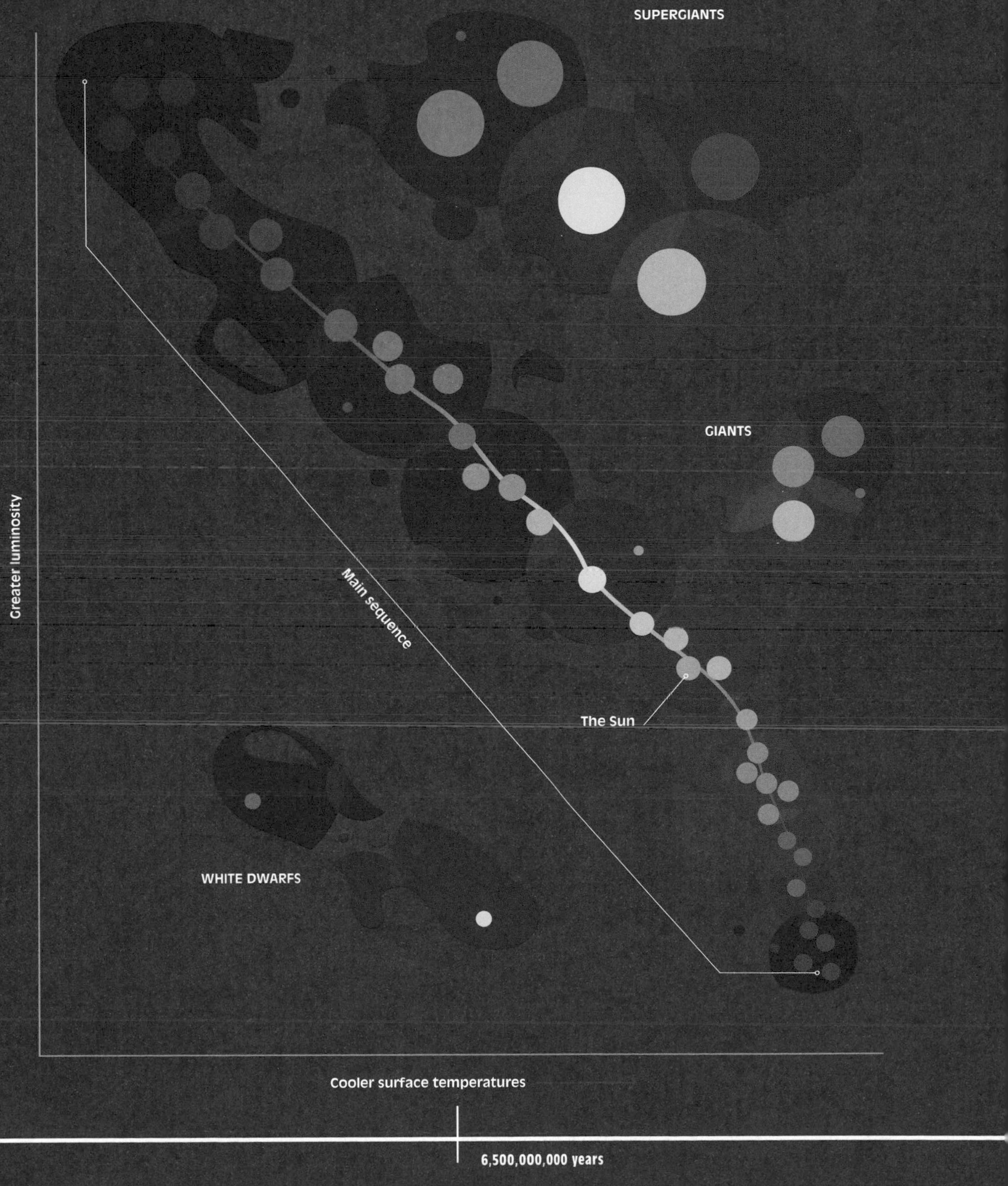

SUPERGIANTS

GIANTS

Greater luminosity

Main sequence

The Sun

WHITE DWARFS

Cooler surface temperatures

6,500,000,000 years

Stellar Birth and Ageing

Stars – just like people – are born, grow up and change as they get older.

Through the study of the Hertzsprung–Russell diagram and the physics of nuclear fusion, astronomers today understand that stars experience very different life cycles depending on how much mass they have when they're born. All stars form when clouds of interstellar gas and dust collapse upon themselves, causing the pressures and temperatures at their cores to get so high that nuclear fusion begins. Stars with greater mass reach greater temperatures and pressures, so their rates of nuclear fusion are higher and their lifetimes are shorter. Stars of lower mass shine more faintly, but can last much longer.

Low-mass stars slowly fuse hydrogen into helium over a long lifetime – so long that the universe is not yet old enough for us to see what the end stage of such a star is like. Physical calculations show that after hundreds of billions of years, the helium that has been produced in their cores becomes so dense that it chokes off the fusion process. The star's outer layers collapse on to the helium core, creating a white dwarf about the size of an Earth-like planet that glows with leftover heat and fades slowly to darkness.

An intermediate-mass star like the Sun can sustain fusion via the proton–proton chain for about 10 billion years before the accumulation of helium in their centres starts to interfere. Imagine a car that never has its oil changed – sludge builds up in its engine; in the star's case, helium builds up in its core, clogging up the usual fusion process. Gravity takes over, and the star begins to collapse upon itself again. Fresh hydrogen piles on to the outer layers of the helium core, creating a shell of hydrogen that violently ignites into even more rapid fusion. The star swells from the new burst of energy, growing in size and luminosity to a red giant.

For every brilliant O star that is born in the universe today, more than 100,000 diminutive M stars are born. That one O star, however, easily outshines all those M stars; its light is fleeting, though, as O stars fuse their hydrogen fuel into helium so quickly that they die before an M star has experienced a millionth of its lifetime.

As nuclear fusion ignites at the centre of an infant star, the outer regions of the gas cloud that formed it are still falling inwards. That material interacts with the fusion energy flowing out through the cloud; narrow jets of material shoot outwards from the star, while the increasing outward pressure blows away excess gas, leaving a thin disc of material orbiting around the star's equator.

Objects that form like stars but have too little mass to sustain nuclear fusion are called brown dwarfs. They weigh only a few per cent of the Sun's weight, and can sometimes fuse hydrogen into deuterium for a short period of time; soon after they form, though, they settle into a slow process of cooling off from the heat of their birth, barely visible by the gentle infrared light they emit.

BIOGRAPHY

English astrophysicist **Arthur Stanley Eddington** (1882–1944) helped to show that the internal structure of white dwarfs is supported by electrons pushing against one another as their atoms are pressed closer and closer together by gravity. The Pakistani-born American astrophysicist **Subrahmanyan Chandrasekhar** (1910–1995) subsequently showed that the electron degeneracy pressure had a limit, and that above about one and a half times the mass of the Sun, a white dwarf would collapse catastrophically into something much smaller. Years later, such compact objects – neutron stars and black holes – were discovered.

Smaller interstellar clouds form low-mass stars that end their lives as white dwarfs. Larger clouds can form intermediate-mass stars like the Sun, which undergo a red giant phase before also becoming white dwarfs.

Low-mass star

White dwarf

Interstellar cloud

intermediate-mass star

Red giant

Interstellar cloud

7,500,000,000 years

Stellar Nucleosynthesis and Reproduction

Even as stars are sending their light outwards throughout the universe, they are manufacturing the universe's elements inside themselves.

The red giant phase is short-lived compared to the main-sequence phase, lasting at most about a billion years. At the end of it, however, the helium core has become so massive, dense and hot that a new process of nucleosynthesis is now possible: the fusion of helium into carbon. At the critical moment, a huge flash of helium fusion occurs, infusing the star with tremendous energy and supporting the star from collapse for a while longer. Several hundred million years later, a series of additional, smaller helium flashes finish off the star completely; the explosions hurl the star's outer layers into space, and the core of helium and carbon becomes a white dwarf.

The creation of carbon is the first of many steps that intermediate and high-mass stars undergo at the end stages of their lives to make more and more different kinds of elements. While it is a red giant, the atomic nuclei within the huge borders of the swollen star are bathed by a steady flow of neutrons flowing outwards from its centre. Over thousands of years, if those neutrons come in contact with the nuclei, they can fuse together to form new elements in a slow but steady chain of growth.

In high-mass stars, nucleosynthesis is far more dramatic. The helium flash that creates carbon is still not powerful enough to hold the star up against gravitational collapse. As more matter falls into the stellar core, temperatures and pressures and densities build to ever higher levels until even more complicated fusion chains commence. Helium fuses into oxygen. Carbon fuses to form neon and magnesium. Oxygen fuses into silicon and sulphur. The processes accelerate madly as the high-mass star continues to race between collapse and the fusion energy that arrests it. Finally, at temperatures of billions of degrees, iron is fused – a limiting process, after which no additional energy can be produced from fusion.

Days later, the star explodes violently as a supernova. The freshly fused elements are blasted outwards into space, bombarded by huge numbers of fast-moving neutrons. Some of the neutrons fuse with the nuclei, forming vast numbers of elements within a fraction of a second. This rapid process, along with the slow process in red giant stars, produces all the elements on the periodic table we know today.

Scattered into the expanse of interstellar space, these elements travel for millions of years as they gradually slow down, coming into contact with other free-floating atoms. Over millions more years, the atoms gather into interstellar clouds and cool slowly to about –260°C (–440°F), just 10 or 20 degrees above absolute zero. The cold clouds then start their own process of collapse, where millions of years later still, a new generation of stars is born.

BIOGRAPHY

In 1954, four prominent English and American astronomers and physicists – **Geoffrey Burbidge** (1925–2010), **Margaret Burbidge** (1919–2020), **William Fowler** (1911–1995) and **Fred Hoyle** (1915–2001) – synthesized past research results with new observations and theories in a treatise, 'Synthesis of the Elements in Stars'. Their work helped lead to our modern understanding that stellar nucleosynthesis is the origin of almost every atom in the universe today that is heavier than hydrogen and helium. Decades later, Fowler was awarded the Nobel Prize in Physics, and Margaret Burbidge became the Inaugural Fellow of the American Astronomical Society.

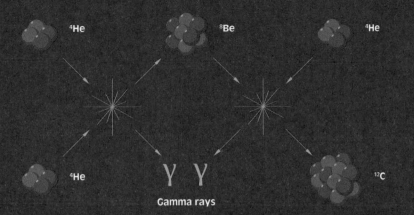

⁴He   ⁸Be   ⁴He

⁴He

Gamma rays

¹²C

Three helium-4 nuclei fusing together to form a carbon-12 nucleus is known as the triple-alpha process, so named because helium-4 nuclei are also called alpha particles. It only occurs at temperatures exceeding 100 million degrees and extreme densities like those found in the hearts of stars.

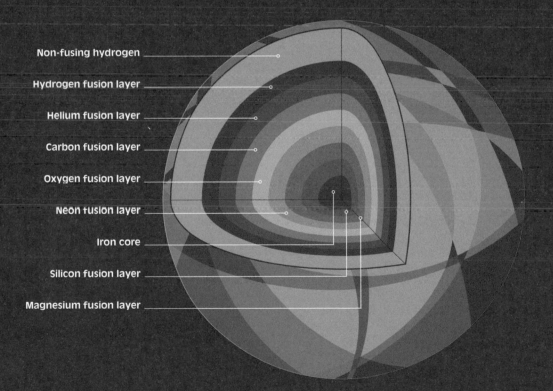

Non-fusing hydrogen

Hydrogen fusion layer

Helium fusion layer

Carbon fusion layer

Oxygen fusion layer

Neon fusion layer

Iron core

Silicon fusion layer

Magnesium fusion layer

In high-mass stars, successive stages of nuclear fusion create numerous layers of newly produced elements, which are blasted into deep space when the stars detonate as supernovae.

8,500,000,000 years

Marking Cosmic Time

The Sun is born.

In the billions of years since the formation of the Milky Way, successive generations of stars lived and died, with each generation spreading its nucleosynthesized elements further and further into every corner of the galaxy. That means the ratio of heavy to light elements in a star is one indicator of that star's age. Stars more recently born have a higher percentage of these elements, whereas stars formed early on have a lower percentage.

Some of those heavy elements, however, don't make it back into new stars. Instead, when a cloud of collapsing gas forms a new star, some of the surrounding material settles into orbit around the star rather than falling into it. If the conditions are right, that excess matter can then come together to form larger bodies – first tiny grains, then pebbles, then boulders, then planetoids. The heavy elements become part of those solid bodies, forming piles of gravitationally-linked rubble, and in some cases cementing together both physically and chemically to create rock.

Furthermore, many heavy elements produced by nuclear fusion immediately start to break down into lighter constituent pieces. By this process of radioactive decay, the numbers of protons and neutrons in these elements' nuclei gradually change, following a predictable pattern of time.

The measurable and reliable nature of radioactive decay means that scientists can deduce the ages of objects in the universe today. For example, if a piece of rock has not been chemically altered since it was formed, we can measure in the rock the amount of a given radioactive element, as well as the amount of by-product elements produced by that element's decay; comparing the two quantities, we can calculate how long ago that rock was formed.

The presence of measurable radioactive decay in rocks was a game changer for the measurement of time's passage in the universe. Until this moment, 'where' something 'was' in the history of the universe depended on the reference point of the Big Bang. Although we have scientifically measured when the Big Bang occurred, nearly 13.8 billion years ago, some uncertainty in that figure will always persist because nobody was there to witness it. Radiometric dating provides a new and much more tangible perspective: henceforth, 'where' in history can be how long ago it happened before the present moment. Even though distant measurements will still be imprecise, we can now hitch them to our own place in time.

It is by using radiometric dating of the oldest meteorites – rocks that formed as the Sun was born – that astronomers have confirmed the age of the Sun. To the precision of current measurements, the Sun is 4,570,000,000 years old.

BIOGRAPHY

American chemical physicist **Bertram Borden Boltwood** (1870–1927) was the first person to publish a study of the age of rocks using the radioactive decay of uranium-238. He measured a range of ages between 400 million and 2,200 million years. A few years later in 1913, English geologist **Arthur Holmes** (1890–1965) published his ground-breaking book *The Age of the Earth*, in which he showed his own results that improved Boltwood's work and solidified the field of radiometric dating.

Every element has one or more isotopes – different versions of the same element with different numbers of neutrons in its nuclei. One well-known use of radiometric dating is with carbon-14, a rare isotope of carbon that is continually replenished in living things until they die. By comparing the ratio of carbon-14 to carbon-12 (the most common carbon isotope) in a sample, it is possible to measure the ages of plant material and animal remains up to about 50,000 years old.

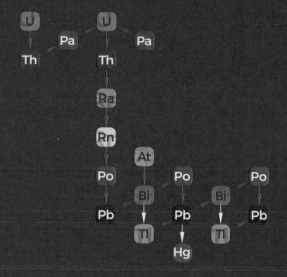

The radioactive decay chain of
uranium-238 consists of a series of
alpha decays (where the nucleus loses
a helium nucleus) and beta decays
(where the nucleus loses a neutron and
gains a proton). One gram of U-238,
after a half-life of 4,468,000,000 years,
will continuously decay until only half a
gram remains; the rest will have decayed
into lead (Pb-206) and other by-products.

Using radiometric dating, astronomers
have calculated that our Sun was born
4,570,000,000 years ago. At birth, it was
surrounded by a swirling disc of dusty
gaseous material, within which the planets
of our solar system would eventually form.

From this time forward, the ages of objects
and matter can be calibrated to the present
day with radiometric methods.

4,600,000,000 *years ago*

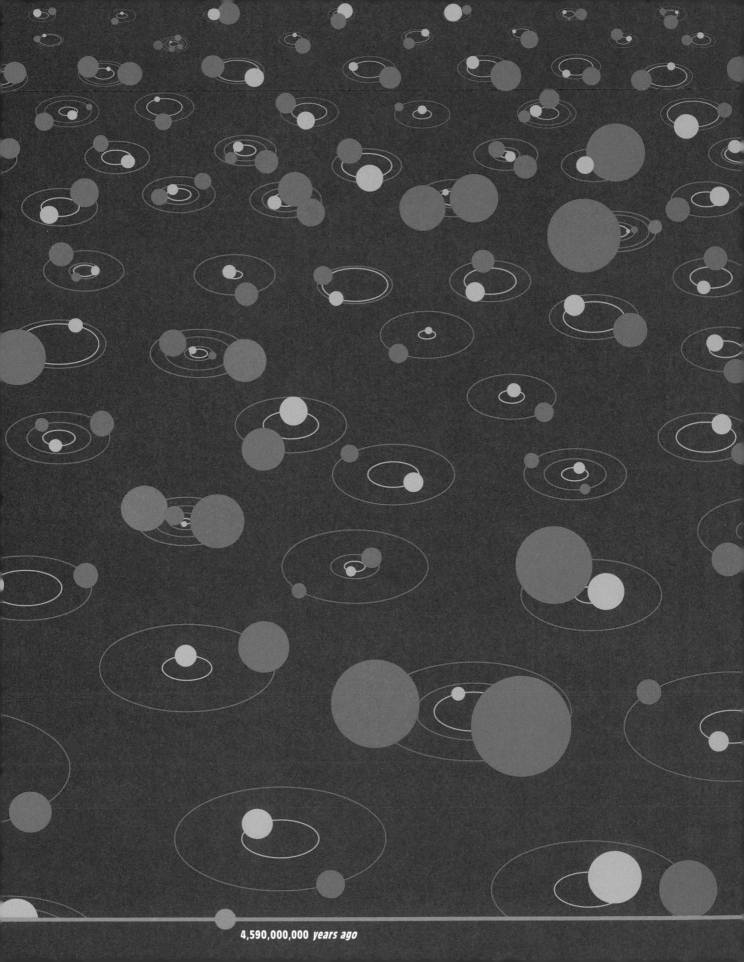

4,590,000,000 *years ago*

7

4.56 Billion Years Ago

The Birth of the Earth and the Moon

On 18 February 2021, a doughty little spacecraft entered the thin atmosphere of Mars.

Over the course of seven minutes, it used a parachute and a rocket-powered sky crane to slow down from nearly 20,000 kilometres per hour (12,400 miles per hour) to human walking pace. The *Perseverance* rover, a car-sized robotic vehicle with 23 cameras and a suite of scientific instruments, touched softly down on four wheels and sent a signal back to its breathless creators: it was ready to explore another world.

Planets, as far as we know, are a dime a dozen in the universe. For almost the entire history of our species, though, we knew of fewer than 10 planets: the ones that orbit the Sun. With the unaided eye, we saw Mercury, Venus, Mars, Jupiter and Saturn – at least, that's what people on the European continent called them. (In Chinese culture, they were the water star, the gold star, the fire star, the wood star and the earth star.) Unlike the Sun, Moon and stars, whose motions cycled predictably over the days and years, these celestial objects wandered across the sky in meandering paths. In more superstitious ancient cultures they were granted wide-ranging powers and even divine status – an incorrect but perhaps understandable conclusion, given their beauty and mystery as inscrutable, unblinking sentinels in the night sky.

Ultimately, scientific study of the planets showed they were completely natural objects – the result of the laws of physics, the birth of the Sun and the passage of time. In the first 10 million years of the Sun's life, it reshaped the vast cloud of gas and dust that had collapsed to form it into a spinning disc of leftover material extending billions of kilometres outwards. Within that disc, tiny particles formed – and from them, pebbles, then boulders, and then planetoids. As time went on, the most massive bodies exerted the strongest gravity, attracting their smaller neighbours to join together with them to build planets large and small. Objects that stayed free from the planets' gravitational influence became comets and asteroids, and many of them still orbit the Sun, nearly unchanged from when they formed more than 4.5 billion years ago.

Looking up in the sky at night, it may seem that not much is changing out there. That sense of eternal cosmic calm might hold on human lifetimes, but the solar system and its denizens are always on the move. Over millions of years, a multitude of small gravitational interactions combine to change the orbits of even the largest of planets; dynamical studies suggest that there might even once have been an additional planet that was gravitationally flung out of the solar system, settling the other planets into the stable orbits in which they travel today. We owe our continued existence to that stability; and we know that planets themselves age and change over time, too, and often drastically. So as we study our planetary neighbour with the help of *Perseverance*, we are studying ourselves as well – and what our ancient history might have been, and what our planet's destiny may be.

Interstellar and Interplanetary Matter

As with any recipe, the making of our planet began with its ingredients.

As interstellar matter throughout the universe coalesced into galaxies, individual clouds of gas and dust collapsed in on themselves to form the first stars in those galaxies. In our Milky Way, the loose interstellar matter that wasn't locked up in stars and black holes included both unprocessed gas left over from the Big Bang – hydrogen and helium – and particles transformed by nucleosynthesis inside stars and thrown back out into interstellar space – all the elements and isotopes from lithium to lanthanum to uranium.

Depending on where and how densely the 'interstellar medium' collects, it can be lit up from outside or inside, producing some of the most beautiful astronomical phenomena in the night sky. The gas and dust right in the vicinity of newly born stars form what are called protoplanetary nebulae – the birthplaces of planets.

Some of the most colourful nebulae result from the endpoints of stellar evolution. Stars like the Sun eject planetary nebulae (so misnamed because they looked like planets to the nineteenth-century astronomers peering through their small telescopes) revealing their white dwarf cores inside. Very massive stars explode as supernovae, creating gaseous remnants glowing with the power of hypervelocity shock waves and remnant radioactive decay.

Reflection nebulae usually shine with bright blue light scattered off grains of dust from stellar light sources nearby. Dark nebulae such as Barnard 68 and the Coalsack are cold opaque clouds that let very little visible light through – but often glow with infrared light produced by the newly born stars that have just formed inside them.

Interstellar chemistry often occurs in the warm regions in and around stars, as well as in giant, cold interstellar clouds just a dozen degrees or so above absolute zero. Simple molecules such as hydrogen (H_2) and carbon monoxide (CO) are the most common; vast numbers of complex chemicals float around in space, too, such as alcohol and antifreeze molecules, as well as big spherical carbon constructions called 'buckyballs'.

How molecules form in space remains a major topic of modern astrophysical research. Atoms can't just run into one another and combine the way they do in laboratories on Earth – they're spread too thin and they move too fast. Atoms could land on grains of space dust, slowly interact to make molecules and then drift away back into space; how could the dust grains have formed in the first place, though, if molecules didn't exist already to build them?

BIOGRAPHY

American astronomers **Edward 'E. E.' Barnard** (1857–1923) and his niece **Mary Calvert** (1884–1974) were perhaps the foremost astrophotographers of their day. Working at Yerkes Observatory in Wisconsin, one part of their work with Yerkes' director **Edwin Frost** (1866–1935) was published in 1927 as a catalogue of 'dark' objects in the night sky. The majority of them were shown to be thick clouds of interstellar gas.

The Orion Nebula, visible from both Earth's northern and southern hemispheres, is a stellar nursery rich with the dust and gas that clump together to build new stars.

4,580,000,000 *years ago*

The Young Sun and Solar Nebula

When it came to the birth of planet Earth, our destiny lay squarely in the details of the Sun's earliest years.

Although the gas cloud in which our parent star formed had an irregular structure, shaped by gravity and rotation, a fair portion of it had probably already settled into a thin, spiralling disc billions of kilometres across. When nuclear fusion ignited in the Sun, the resulting heat and radiation ploughed outwards in every direction; the sparser regions of the nebula were blown away back into deep space, while the disc remained.

The surviving disc wasn't quiet during this time, either. As the new Sun's heat and solar wind coursed through the disc material, gravity and electromagnetic forces shaped it into denser and sparser bands, while vast sheets of lightning millions of kilometres across may have flashed all around. Lighter gas particles such as hydrogen and helium were pushed more than half a billion kilometres back from the Sun. Rarer and denser grains of rock and metal stayed closer in.

Millions of years later, although prodigious amounts of energy continued to emanate from the Sun (and still do today), much of the initial tumult had died down. The ring-like regions of the protoplanetary disc – spinning streams of gas and dust – started to coalesce into larger bodies. At first they might have started sticking together due to static electricity – the same kind of effect that keeps loose hairs clinging to your jumper on a dry winter's day. At some point gravity took over, and the assembly of the planets began in earnest.

The very young star HL Tauri has a protostellar nebula and disc around it that appears to be similar to the kind the Sun had around itself 4.57 billion years ago. In this picture taken with the Hubble Space Telescope, a jet-like cloud glows with the energy from escaping stellar radiation, while a bright blue nebula blocks our view of the interior. The inset image, taken with the Atacama Large Millimeter Array (ALMA), gets through the glare by looking at radio waves coming from inside the cloud; it shows that the star itself is still shrouded in glowing gas, while rings are starting to coalesce in the disc that surrounds it – the sites of new planet formation.

BIOGRAPHY

The Swedish philosopher **Emanuel Swedenborg** (1688–1772) and German philosopher **Immanuel Kant** (1724–1804) were early proposers of the idea that the Sun and planets formed within a disc-like solar nebula. The French physicist and mathematician **Pierre-Simon Laplace** (1749–1827), who first developed the field of celestial mechanics, developed the nebular hypothesis within a mathematical framework that forms the basis of the modern scientific picture.

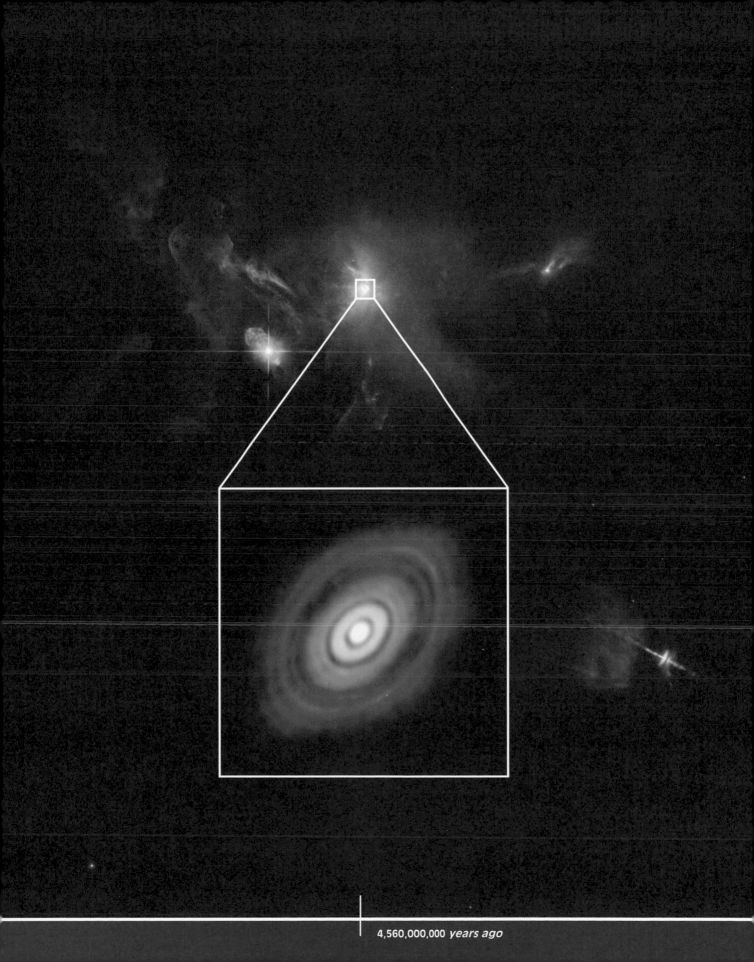

4,560,000,000 *years ago*

The Planets

Earth formed as one of a large family of planets.

Each ring of material that formed a planet within the solar nebula had within itself a different density, a different total mass and a different mixture of elements available. With unique ingredients, each planet turned out to be unique, while also having similarities with its fellows.

The inner planets of the solar system have cores of metal – almost all iron and nickel – and crusts of rock that are made mostly of silicon and oxygen compounds. These heavy atomic elements were the ones that were able to withstand the Sun's punishing winds of hot charged particles and stay intact within 250 million kilometres (150 million miles) of their host star; however, these elements comprised less than 1 per cent of the atoms in the protoplanetary nebula. Not surprisingly, they are much smaller than the other planets, and their gaseous atmospheres are far outweighed by their solid and liquid planetary components.
• Mercury cannot sustain an atmosphere because of its close proximity to the Sun. Its surface shows the remnants of powerful geology-shaping impacts and vast volcanic flows that appear to have lasted for a billion years after the planet first formed.
• Venus might have had a geological history much like our own planet. The thick carbon dioxide atmosphere it generated around itself, however, was so good at trapping heat that its surface temperature grew hotter than even that of Sun-baked Mercury.
• Earth has just enough of a nitrogen–oxygen atmosphere to stabilize its surface temperature around the freezing point of water. Not too hot and not too cold, its surface is covered with sloshing oceans, drifting continents and carbon-based life.
• Mars might have had rainclouds, rivers and seas about a billion years after it formed. But soon afterwards it ran out of heat; today, a wispy-thin atmosphere and a frozen, desiccated landscape provide tantalizing clues to the possible existence of once-living organisms.

The outer planets of the solar system have cores of rock, each of which is more than twice the mass of all the inner planets put together. The gas particles that were blown away from the Sun stayed in the solar system and collected at the distance of the outer planets' orbits, becoming the main component of these gaseous giants. They also have no solid surface; at the bottom of their atmospheres, the pressure is so high that hydrogen gas is compressed into a thick layer of metallic liquid.
• Jupiter has double the mass of all the other planets in the solar system put together. Its Great Red Spot is a storm larger than Earth that has raged for more than four centuries.
• Saturn, in turn, has double the mass of all the planets other than Jupiter combined. Its remarkable ring system is comprised of trillions of bits of ice and rock, and is expected to persist for about 100 million years before it dissipates into interplanetary space.
• Uranus and Neptune are each about 20 times the mass of Earth – half of that in the core and half in the atmosphere. Like Jupiter and Saturn, these gas giants have numerous moons in orbit around them – miniature planetary systems unto themselves.

The dwarf planets are indeed small – none currently known are larger in diameter than the Moon – yet are nevertheless fascinating worlds that orbit the Sun. Many of them – such as Ceres, discovered in 1801 by Italian Giuseppe Piazzi (1746–1826), and Pluto, found in 1930 by American Clyde Tombaugh (1906–1997) – were classified as planets when they were first discovered, then reclassified after scientific study showed them to be tiny.

Mercury
Diameter: 4,880 km/3,030 miles

Venus
Diameter: 12,100 km/7,520 miles

Earth
Diameter: 12,760 km/7,930 miles

Mars
Diameter: 6,790 km/4,220 miles

Jupiter
Diameter: 143,000 km/88,855 miles

Saturn
Diameter: 121,000 km/75,185 miles
(excluding rings)

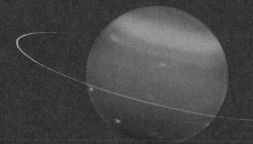

Uranus
Diameter: 51,100 km/31,750 miles

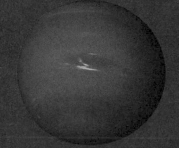

Neptune
Diameter: 49,500 km/30,760 miles

4,540,000,000 *years ago*

Moons, Asteroids, Comets and Collisions

Based on geological and astrophysical evidence, planet formation was an energetic and unruly process.

Numerous protoplanets existed throughout the nascent solar system; almost all of them formed around the same time, and they frequently crashed into one another as they orbited around the Sun with different directions and speeds. After 50–100 million years, the largest objects had collected the majority of their smaller kin, forming the cores of the planets we know today; that process of planet building led to all manner of collisions large and small, including titanic impacts that reshaped entire worlds.

Around 4.51 billion years ago, a Mars-sized protoplanet (astronomers have nicknamed it 'Theia') crashed into the infant Earth at a glancing angle. Imagine a ripe melon striking a pulpy pumpkin, not quite head-on, at thousands of kilometres per hour; both objects shattered of course, and some pieces were hurled far away into the solar system, but their mutual gravity pulled most of their combined matter back together into a single body. About 1 per cent of the material remained in orbit around the newly enhanced planet, and that little remainder coalesced over time to form the Moon.

By the time the major planets formed, only a tiny fraction of the solar nebula remained unincorporated into large bodies. Ranging in size up to a few kilometres or so across, amalgamations made primarily of ice and rock became comets, while collections of mostly rock and metal became asteroids. Although all their combined mass is less than that of Earth's moon, there are millions of them still zipping around the solar system today, waiting for their chance – if it ever comes – to add their weight to one of their bigger siblings.

Meteorites are pieces of solar system material that have fallen to Earth and survived both the journey from outer space and the environmental effects after they arrived. They contain material that has often been unchanged for billions of years – historical artefacts that have preserved the ancient history of the solar system.

The *NEAR-Shoemaker* spacecraft, named after the pioneering American planetary scientist Eugene Shoemaker (1928–1997), studied the saddle-shaped asteroid 433 Eros in orbit from 14 February 2000 to 12 February 2001, and then ended its mission by landing on the 34 kilometre-/21 mile-long asteroid.

In 1976, the Canadian-American and American astrophysicists Alastair G. W. Cameron (1925–2005) and William Ward (1944–2018) proposed that the Moon was formed when a Mars-sized protoplanet crashed at an angle into the molten Earth about 100 million years after the birth of the Sun. A decade later, Cameron and Swiss astrophysicist Willy Benz (b. 1955) produced the first supercomputer simulations of how such a collision could have resulted in a proto-Moon 4.5 billion years ago.

In 1994, the comet Shoemaker-Levy 9 broke into pieces under the influence of Jupiter's gravity and crashed over a series of days into Jupiter's atmosphere. Each of the seven largest impacts released more energy than all of Earth's nuclear and conventional explosives combined, and created temporary dark holes thousands of kilometres wide.

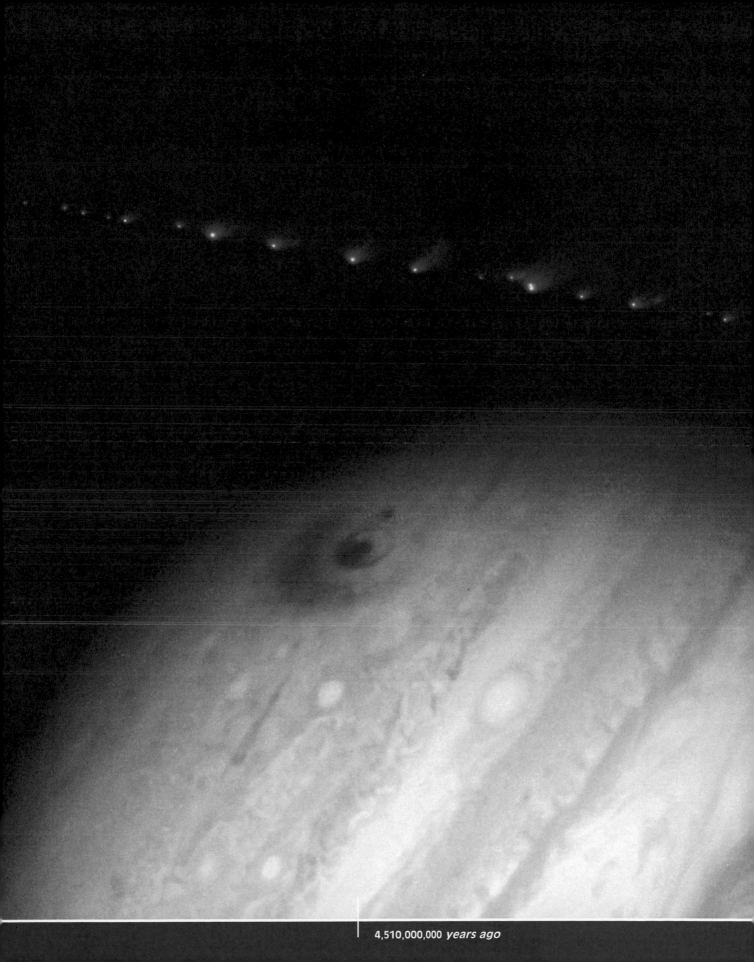

4,510,000,000 *years ago*

Planetary Systems and Exoplanets

Just a generation ago, astronomers could only guess if any planets at all existed outside our solar system.

It made sense that many should exist; after all, the laws of physics that governed the formation of planets around the Sun hold true throughout the universe, so every new star stands a chance of producing its own system of planets, moons, asteroids and comets. The challenge was finding them.

Impressively, astronomers in recent years have used a handful of techniques to find exoplanets – planets beyond our solar system that orbit other stars. The Doppler method measures the tiny wobbling motions of a star to detect the presence of planets that tug back and forth on the star as they orbit. The transit method looks for periodic dips in the brightness of a star that would indicate that a planet is passing in front of the star as it orbits. Perhaps the most challenging method is direct imaging; unless the host star's light is almost completely blocked out while the picture is being taken, the tiny flicker of the planet's heat and light will be overwhelmed and undetectable.

Due to the limitations of our current telescope technology, only one small corner of the Milky Way has been searched for exoplanets. Even so, more than 4,000 worlds have thus far been discovered! From these early results, astronomers predict that there are many billions of planets in our galaxy alone.

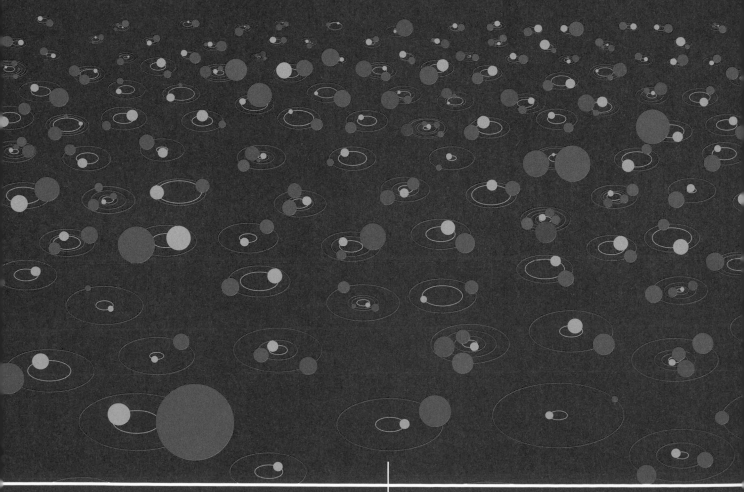

The distance between Earth and Pluto is a breathtaking 5 billion kilometres (3 billion miles), and it took NASA's *New Horizons* space probe nearly a decade to get there. The nearest known star to the Sun, on the other hand, is 40 trillion kilometres (25 trillion miles) away – 8,000 times more distant. Getting to a planet orbiting that stellar neighbour would take a prohibitively long time without a revolutionary advance in space travel.

BIOGRAPHY

In 1995, Swiss astronomers **Michel Mayor** (b. 1942) and **Didier Queloz** (b. 1966) discovered the first planet orbiting a Sun-like star other than our own by tracing the influence of its gravitational pull on the motion of its host star 51 Pegasi. For their work, which opened the floodgates of exoplanetary discovery, they were awarded the 2019 Nobel Prize in Physics.

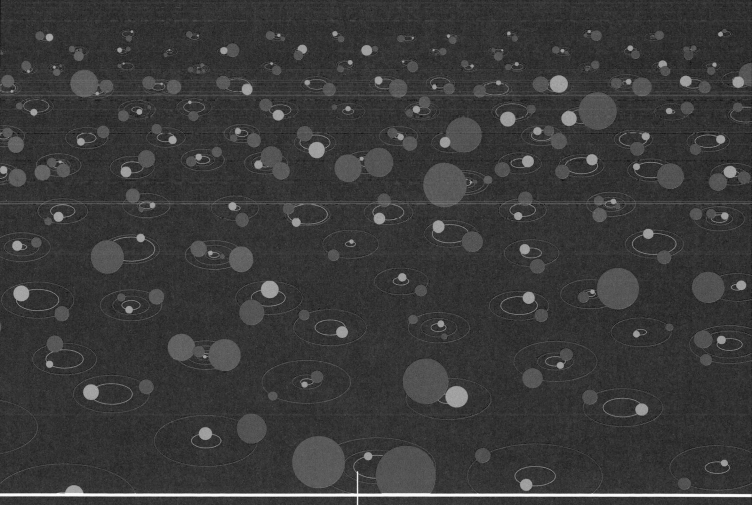

4,400,000,000 *years ago*

3,900,000,000 *years ago*

8

3.8 Billion
Years Ago

Birth of Life on Earth

For 10 billion years, we have watched the universe evolve.

Once a blazing-hot crucible, the cosmos cooled until it was a hospitable environment for individual objects to form inside it. In vast gaseous nurseries, stars were born; they aged, reproduced and died. They spent their lives in vast galactic ecosystems that varied widely in size, shape and activity. Planetary systems formed, enveloped in the environments produced by their host stars; within them planets were born, each the product of its composition and surroundings, all slowly cooling as they stabilized and evolved.

Now, on Earth, as the planet itself ages and evolves, new objects are forming with a remarkable property: the ability to preserve and communicate so much information that they can reproduce themselves almost exactly, just by carefully processing the available resources around them. Certain environmental conditions on Earth made these objects possible: the right mix of light and heavy elements, steady warm temperatures and lots of liquid water. Over billions of years, these carbon-based reproducers will evolve into millions of varieties and spread across the Earth, fundamentally altering and shaping their planet.

Life as we know it is unusually adept at both accurate reproduction and adaptation to new environments – so good, in fact, that we tend to assume it must be uniquely special in the universe. We have already seen enough of cosmic history, though, to know that natural processes following the same basic physical rules can result in dazzlingly beautiful and complex phenomena of many kinds, including some that would have been unimaginable had we not actually observed their existence. Does life on Earth fall into such a category – amazing for sure, yet in a cosmic sense completely ordinary?

To answer that question, it would certainly help if we could find another example of a planet on which organisms like us exist. The technology to find extraterrestrial life isn't quite there yet, but we're close; among the thousands of planets astronomers have found in the past quarter of a century, a handful seem to be quite similar to Earth. In a few decades, our telescopes and detectors may be able to show definitively that one or more of them are home to things we would recognize as being alive.

Until then, we have plenty of mysteries to solve when it comes to our own cosmic genealogy. As just one example, stars evolve over eons, and each new generation can take millions of millennia to mature; some organisms on Earth, by contrast, can have great-grandchildren in less than an hour. What were the physical and environmental conditions billions of years ago that led to the development of such efficient reproduction? More profoundly, is a complex life form like a human being a single organism or a composite of many organisms – or even, indeed, an entire ecosystem? Fundamentally, a human may be no 'better' than the microbes from which it evolved, or the trillions of microbes that coexist in and around it at every moment. Then again, a human may be no 'worse' than Earth itself – a body crisscrossed with systems, host to vast numbers of smaller bodies and all following a common set of rules: the laws of nature that have been operating on scales large and small since the birth of the universe.

Earth Solidifies, Differentiates and Evolves

The heat from all of the collisions that brought the pieces of our planet together was tremendous.

Earth was completely molten for millions of years – its rock and metal content blended in a thick, gooey fluid mixture. Imagine a jar filled with water, oil and sand, continually shaken and stirred at super-high temperature and pressure.

Once the bulk of Earth's material had been gathered and the collisions slowed to just a trickle, our planet did just what our jar would do once it were left undisturbed: it began to separate into layers – a process called differentiation. Gravity caused the metal to fall first and furthest, all the way down to Earth's molten core. Rocky material dropped downwards as well, sifting through the body of the world until it was stopped in the mantle, buoyed on top of the dense metals.

This slow-motion falling process released huge amounts of heat outwards through the planet. Earth's surface cooled quickly as its heat flowed out into space. Meanwhile, low-density elements such as hydrogen, carbon and oxygen combined with the rocky and metallic elements (and themselves) to make compounds and minerals. Where the temperature dropped enough, the liquid material froze to form Earth's crust.

The process of Earth's cooling and solidification continues steadily to this day. Although the crust was largely solid after about half a billion years, the temperature at Earth's core today, after more than 4 billion years, is still probably more than 5,000°C (9,000°F) – about as hot as the surface of the Sun. Our planet's interior heat continues to shape the surface of our world, as volcanic activity continually spurts magma through cracks and thin spots in the crust, adding new material on to our land and into our atmosphere.

Despite the stifling heat at the centre of the Earth, the innermost iron core is solid. The pressure of the planet pushing down on it is so immense – more than 3 million kilograms per square centimetre, or 42 million pounds per square inch, that the metal is crushed into a super-hot crystalline form. The outer part of the core is liquid, and helps create Earth's magnetic field as it sloshes and spins.

With current technology, humans cannot drill more than a few kilometres into Earth's crust. Geologists thus study the interior structure of our planet by tracing the vibration patterns of earthquakes. To examine what the material deep inside the Earth might look like, planetary scientists also study meteorites that were pieces of asteroid and dwarf planet cores before they fell to Earth.

Earth is more than 12,700 kilometres (13,800 miles) in diameter. Its crust is, however, only 15–20 kilometres (9–12 miles) thick on average, and comprises less than 1 per cent of our planet's volume. Just a kilometre or two down, the crust already reaches temperatures exceeding 65°C (150°F).

BIOGRAPHY

The Croatian geophysicist **Andrija Mohorovičić** (1857–1936) studied tornadoes early in his career. In 1909 he examined seismographic measurements of an earthquake centred near his home and concluded that Earth's solid crust rests on a mantle that is sharply different in structure and composition. The boundary between crust and mantle is known today as the Mohorovičić discontinuity (or 'Moho' for short) in his honour.

Inner core

Outer core

Mantle

Crust

3,800,000,000 *years ago*

Oceans and Continents

Earth is constantly evolving, although not in the same way that living species do.

Although our planet is made mostly of solid rock and metal, the extreme temperatures and pressures inside it cause that dense material to behave in many ways like a fluid. To humans, that motion is achingly slow and almost undetectable. When viewed through the lens of cosmic time, however, Earth is a dynamic and ever-changing place, with a network of interconnected cycles and systems.

The hard, shell-like layer near the surface of any planet or moon is called the lithosphere. Earth's lithosphere, which consists of the crust and the solid portion of the upper mantle, isn't a single piece; rather, it's made of continent-sized slabs of rock that float on the fluid part of the upper mantle (called the asthenosphere) and slide slowly past one another. Imagine a wintry lake covered with sheets and chunks of ice floating on the water; the ice pieces are continually running into one another, pushing against the edges of their neighbours, piling together and pulling apart. That's what Earth's tectonic plates are doing right now, moving a few centimetres per year – about the rate your fingernails grow.

Perhaps the most important system to the development of life on Earth is the water cycle. As our planet formed, water molecules from comets and other icy impactors were trapped within the molten rock; as magma slowly reached the surface over millions of years and cooled down, the water escaped into our atmosphere as vapour, cooled further and then fell back to Earth as rain. Although the total amount of water on Earth is tiny compared to the rock, it was still enough to cover the thinner parts of the crust with a layer of liquid up to several kilometres deep. Oceans and seas cover more than two-thirds of the world today.

The flux of solar energy from the Sun today is several thousand times greater than the flux of energy outwards from Earth's interior. The Sun's influence, however, extends downwards only a short distance into the crust. Thus, while the Sun powers activity near Earth's surface – such as climate and weather, the water cycle and photosynthetic life – we still need the slow and steady heat from Earth's mantle and core to power our dynamic planet.

Processes below the surface such as plate tectonics and volcanoes get their energy from Earth's interior. The differentiation of Earth early in its history into crust, mantle and core released large amounts of gravitational potential energy. The heat left over from the Earth's formation seeps gently outwards from our planet's core; and a similar amount of heat is released by the radioactive decay of elements such as uranium and thorium.

BIOGRAPHY

Scottish geologist **Charles Lyell** (1797–1875) helped found the modern scientific study of Earth. His three-volume work *Principles of Geology* (1830–1833) explained how natural processes slowly and surely change the planet over many thousands and millions of years. This idea of uniformitarianism – in contrast to catastrophism, which invokes drastic events as the cause of significant geological change – is a key pillar of modern science: the laws of nature have operated everywhere throughout the history of the universe in the same ways, producing a cosmos filled with vastly varying conditions yet all following the same basic rules.

German geoscientist **Alfred Wegener** (1880–1930) greatly advanced the study of weather and of Earth's polar regions in his career. He is, however, best known for the theory that continents drift slowly across the surface of Earth, an idea that formed the foundation of modern plate tectonics. Wegener tragically died during a scientific expedition during the harsh winter on Greenland.

3,600,000,000 *years ago*

Origin of Life – Still a Mystery

Billions of years ago, the odds of life as we know it developing on Earth were literally astronomical.

So many things about our planet are just right for supporting life that finding another place just like it nearby appears unlikely indeed. On the other hand, given the size and age of our universe and the vast numbers of galaxies, stars and planets that have existed since the Big Bang, it seems overall quite likely that we are not alone in the cosmos.

Earth is often said to be in a 'Goldilocks zone' – referring to the story of the girl who was the uninvited guest of three bears – where conditions are neither too hot nor too cold, allowing liquid water to exist on its surface. Other cosmic coincidences that have helped the birth of life on our planet include the solar system's relatively calm location in the Milky Way galaxy; the Sun's continuous, gentle warmth;

the location of Jupiter as a gravitational shield; and the stabilizing orbital and tidal influence of the Moon.

The biological origins of life remain scientifically mysterious. Fossilized stromatolites – mats of organic material produced by large colonies of ancient microbes – show evidence of living things on Earth dating back some 3.8 billion years. Fossils, however, are rare and incomplete records of the distant past; very early life forms were fragile structures that most likely have not survived to the present day. What scientists do know is that very soon after the ingredients to support life – steady warmth, liquid water and key chemical elements such as carbon and nitrogen – were readily available on Earth, what we today would recognize as life began to exist.

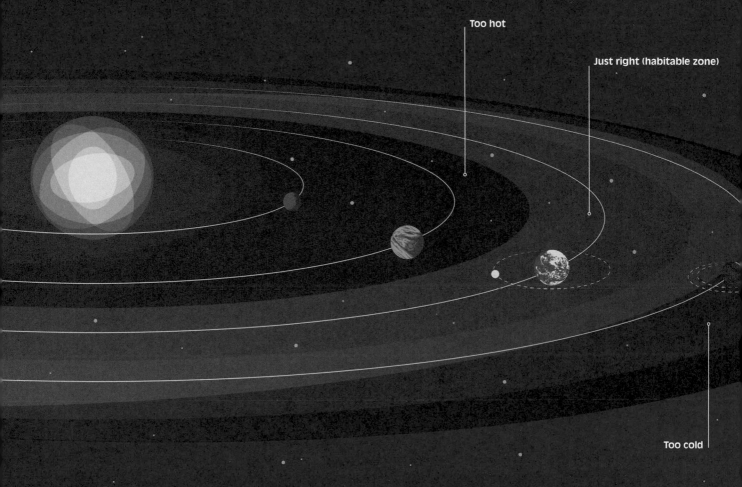

Too hot

Just right (habitable zone)

Too cold

Reproduction is perhaps the most distinctive characteristic of living organisms. The first molecules that could produce proteins as well as store and copy genetic information, albeit inefficiently, were ribonucleic acids or RNA. There may have been a time on Earth when life was so basic that the only genetic activity was based on those compounds – a so-called 'RNA world'. If this is true, astrobiologists think there may exist 'RNA worlds' out in the universe where such very primitive life is the only kind of life on those planets.

DNA, or deoxyribonucleic acid, and RNA are similar, but a key difference in their composition causes DNA molecules to be structured as two long chains of atoms that coil around one another in a twisting spiral. This 'double helix' structure makes the DNA chains stable enough to store large amounts of genetic information and flexible enough to reproduce themselves repeatedly and accurately. DNA is the essential molecular genetic vehicle for all cell-based life on Earth.

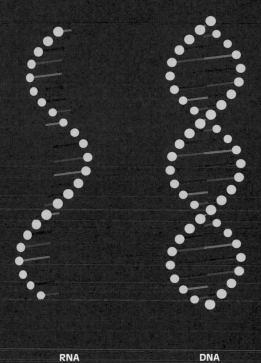

RNA   DNA

BIOGRAPHY

English physical chemist **Rosalind Franklin** (1920–1958) was a pioneer in using the technique of X-ray crystallography to study the microscopic structures of both living and non-living things. Her findings during her brilliant career, cut sadly short by cancer, helped scientists understand the complex molecules essential for life to exist, including nucleic acids such as DNA.

3,250,000,000 *years ago*

Primitive, Microscopic Life

Life's progression from simplicity to complexity happened in ladder-like steps, with each rung taking millions of years to climb.

Strands of DNA molecules probably built basic living things just a few micrometres long, similar to the components that are inside modern-day organisms. A fundamental breakthrough was the development of membranes that could surround and contain numerous different pieces of organic material, allowing them to work together as a single system: a cell.

The earliest cellular life forms to appear were prokaryotes, with their genetic material mostly just floating around within their cell membranes, adapting to a still-volatile Earth. They survived in extreme environments such as volcanic hot springs and undersea vents, as well as in benign freshwater pools and shallow seas. As they grew increasingly complex, a major development occurred about 2 billion years ago when some cells developed a nucleus inside them with its own membrane, where their DNA was contained and organized. Spurred on by this modification, eukaryotes – life forms with these kinds of cells – built even more complicated internal structures, and some grew to many thousands of times larger than prokaryotes.

About a billion years after their first appearance, some eukaryotes began to develop methods to transport chemicals reliably and efficiently not just within a cell, but also to other cells. Evolutionary biologists think that over time, very simple organisms might accidentally have increased their individual chances for survival by working together. Then the cooperating creatures literally combined into systems that took on a life of their own. Groups of cells united into single living systems – and multicellular life was born.

From there, cells further specialized to perform specific functions: to sense their surroundings, to regulate internal conditions and to protect against threats. The large, complex life forms they comprise today – from plants to pets to people – often have more cells in them than there are stars in the universe's largest galaxies.

Endosymbiosis – the process of bringing other life forms together within a single, more complex organism – is thought by scientists to have been a primary force in changing life at the cellular level. Important examples might have been ancient versions of chloroplasts that use sunlight to convert water and carbon dioxide into sugar, and mitochondria that can convert sugars and oxygen into usable energy. Long ago, those two types of cells were likely incorporated into large eukaryotes to enhance those cells' ability to make food and power their life processes. Today, chloroplasts and mitochondria are no longer separate life forms, and are instead organelles inside other cells – powerful parts of a greater whole.

BIOGRAPHY

The American biophysicist and microbiologist **Carl Woese** (1928–2012) pioneered the modern scientific study and categorization of early life on Earth. In a 1990 paper with colleagues **Otto Kandler** and **Mark Wheelis**, he proposed the modern-day classification of life into three domains based on their molecular structures and sequences: archaea, bacteria and eukaryotes.

The 2020 Nobel Prize in Chemistry was awarded to American biochemist **Jennifer Doudna** (b. 1964) and French microbiologist **Emmanuelle Charpentier** (b. 1968) for their development of a method of editing the genetic code of living organisms. CRISPR-Cas9, short for 'clustered regularly interspaced short palindromic repeats-associated protein 9', can be used as genetic 'scissors' to slice out unwanted pieces of DNA – a mechanism that first developed naturally in single-celled prokaryotic life.

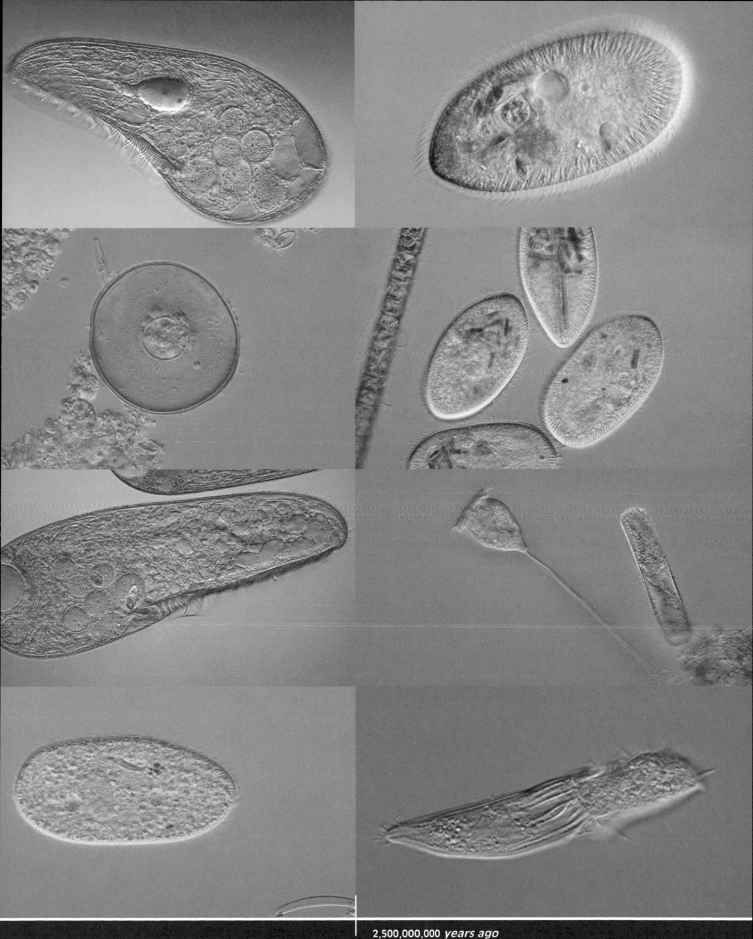

2,500,000,000 *years ago*

Evolution by Natural Selection

Two and a half billion years ago, Earth not only teemed with life; our planet was starting to get crowded.

Pieces of genetic material – mostly RNA and DNA – were rampantly combining all over the world. Most of the billions upon billions of combinations failed to achieve any lasting results, but many others were able to exploit the available physical conditions and chemical processes on our planet to reproduce successfully and rapidly. These were the first members of the genetic domains of life, which subdivided over millions of generations into ever finer gradations of the categories of life: kingdoms, phyla, classes, orders, families, genera and species.

Inevitably, competition appeared between different types of life forms. Those that were able to grow stronger, live longer, reproduce faster or even eliminate their rivals increased their numbers and their presence. As time went on, while single-celled life grew increasingly complex, some organisms appeared that banded together with other, similar organisms to produce results that no one creature could achieve alone; that led first to colonies of bacteria and other single-celled life, and then ultimately to multicellular life forms.

The unending chain of genetic generation continues to this day. Although some of it happens intentionally in laboratories, almost all of the new combinations of RNA and DNA appear on Earth due to random mutations occurring naturally. How does nature select the ones that will persist? Thanks to the continual introduction of new variations, the species that survive – including us humans – must demonstrate a superior ability to compete, cooperate and adapt.

Humanity's fight against infectious disease is a quintessential demonstration of evolution by natural selection. Variations of known viruses and bacteria regularly appear as genetic mutations continually occur. If one variant is able to move efficiently between people, reproduce quickly and defeat efforts to destroy it, that germ can create a worldwide pandemic within just a few weeks. The human species can survive the attack by using cooperative behaviour; first to slow down transmission of the competing life form, and then to develop and deliver vaccines that help individual humans adapt to it with an improved immune system.

The appearance of photosynthetic organisms such as blue-green algae caused the first major evolutionary transformation on Earth after life first appeared. By turning sunlight, water and carbon dioxide into nutrients, they released vast amounts of oxygen – a waste product that proved to be a deadly poison to most living things on Earth at that time. After hundreds of millions of years, new creatures began to adapt to the Great Oxygenation, using the toxic gas to their advantage; and ironically, after 2 billion years of further evolution, almost all life on Earth today needs oxygen to survive.

BIOGRAPHY

English naturalist **Charles Robert Darwin** (1809–1882) began his scientific training as a physician, but opted to work as a geologist instead. During his famous five-year voyage on the research ship HMS *Beagle*, he observed properties of both the Earth and of living things. For two decades he carefully studied the process of evolution by natural selection. When in 1859 he finally published his first book on the subject, *On the Origin of Species by Means of Natural Selection*, it changed the course of biological science forever.

British natural scientist **Alfred Russel Wallace** (1823–1913) was already renowned for trailblazing ecological research when he, like Charles Darwin, concluded that evolution occurs by natural selection. On 1 July 1858, Wallace and Darwin had their theories jointly presented in London. Wallace was a pioneer in natural history exploration, environmental activism and even astrobiology; he was the first biologist to write a book (*Man's Place in the Universe*, 1904) that scientifically examined the possibility of extraterrestrial life.

1,500,000,000 *years ago*

1,400,000,000 *years ago*

9

1.3 Billion Years Ago

Two Black Holes Collide

The first confirmed scientific detection of gravitational waves on Earth.

As carbon-based life took hold on Earth and evolved increasingly complex forms, life went on unabated throughout the universe. In billions of galaxies, trillions of stars were born and sent their light shining into space. Whereas most of those stars are still alive today, the most massive among them exhausted their nuclear fuel rapidly, and as they died their spent cores collapsed to form black holes. Those small but gravitationally powerful stellar remnants orbited in their galaxies, mostly journeying in isolation; stars are usually so far apart compared to their sizes that if they were the size of apples, the nearest neighbour to a star in Scotland would on average be in France.

Very occasionally, though, some of those wandering black holes would indeed encounter another object along their way. Under just the right conditions, they could fall into orbit around one another; and if the other object were also a black hole, they would begin a delicate gravitational dance as a binary black hole system.

Now if gravity were just an attractive force between two objects, as Isaac Newton described it centuries ago, this binary black hole system might have stayed just as it was forever – or at least until a third object intruded upon the pair. But as Albert Einstein explained it in the 1900s, gravity only acts exactly like a force on typical human scales; in actuality, gravity is the curvature of space and time caused by the concentration of matter at different locations in the universe. That subtle difference shows itself in surprising ways – including what could happen to two black holes: their orbital energy slowly dissipates through gravitational radiation. As a result, they could start swirling closer and closer together over millions or even billions of years, until their event horizons merged and the two black holes became one.

About 1.3 billion years ago, just such an event occurred in a distant galaxy. Two black holes, each about 10 million times the mass of the Earth, coalesced in a cataclysmic collision. The shock of the crash created rippling vibrations in space itself – and a thousand times more energy was released in those few moments than the Sun will have produced in its entire 10 billion year-long lifetime. That energy streamed into space not as light but as a gravitational wave that momentarily distorted the very dimensions of length, width and height around it.

The titanic pulse of gravitational energy travel started its journey when the most evolved carbon-based creature on Earth was a fungus. By the time it finally reached us, the wave had lost so much strength that it was only straining the structure of space by less than one-thousandth the width of a proton. Yet despite that tiny amplitude, scientists listening with a remarkable experimental system called LIGO (The Laser Interferometry Gravitational-Wave Observatory) heard its unmistakable chirp on 14 September 2015 as it passed through our planet. The event was called GW150914 – a prosaic name that belied its scientific significance. At that moment, the era of multi-messenger astronomy began – and a wholly new way for humans to sense the universe came of age.

When two black holes combine, the collision sends shock waves through space-time, briefly and almost imperceptibly distorting the shapes of faraway moons and planets.

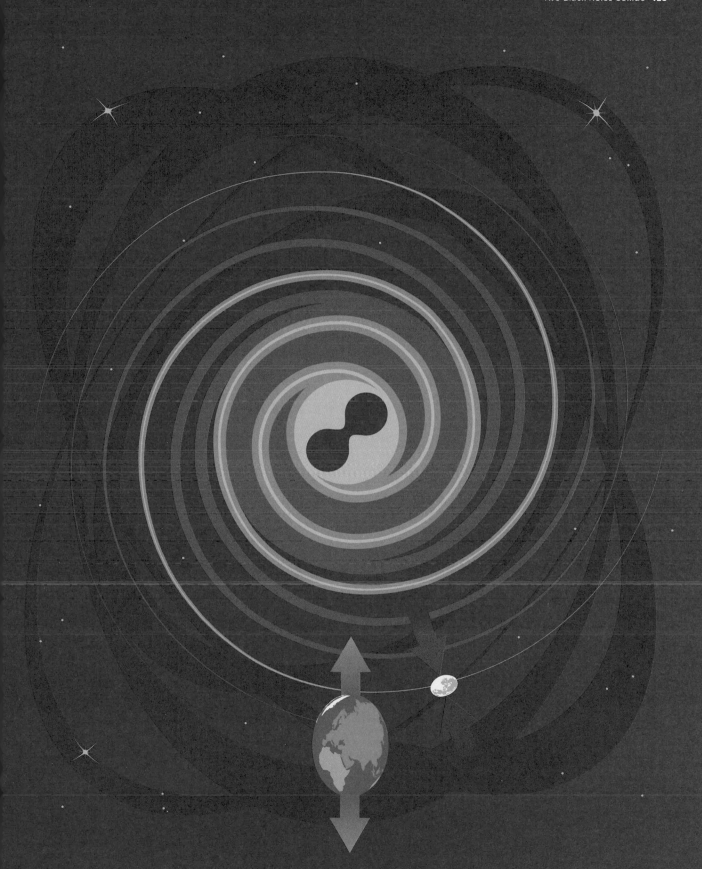

Gravity

Once the cosmic microwave background radiation was established 380,000 years after the Big Bang, gravity began to organize the matter in the universe.

Perhaps surprisingly, gravity is far weaker than the other fundamental (strong nuclear, weak nuclear and electromagnetic) forces – just the tiny bit of static electricity on a balloon, for example, can cause your hair to defy the gravity of the entire Earth. On the other hand, the nuclear forces only work at subatomic distances, and electromagnetic force has both positive and negative versions that usually cancel out one another. So, gravity controls almost all the motion of terrestrial and celestial objects at interplanetary, interstellar and intergalactic scales.

Gravity's acceleration of massive objects changes not only their locations but also their shapes. Tides are caused when one side of a large object is more strongly pulled by gravity than the other, stretching and shearing the object's internal structure. Over billions of years, the tides caused on our planet by the Moon and the Sun have gently pulled ocean waters back and forth across Earth's shorelines, creating conditions that helped lead to the evolution of carbon-based life; at the same time, the tidal influence on Earth's liquid core has released internal heat that has helped to warm those oceans from below.

BIOGRAPHY

The English physicist and mathematician **Isaac Newton** (1642–1727) had to interrupt his university education because of a bubonic plague epidemic. While studying at home during those two years, Newton deduced the mathematical origins of calculus, the law of universal gravitation, and his three eponymous laws of motion – although he did not publish most of this work until many years later. During his lifetime, Newton often recounted a story of being inspired to explain gravity after he saw an apple fall from a tree at his home.

1,400,000,000 *years ago*

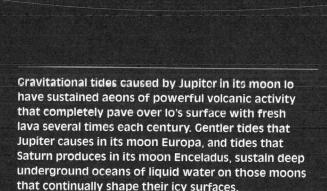

Gravitational tides caused by Jupiter in its moon Io have sustained aeons of powerful volcanic activity that completely pave over Io's surface with fresh lava several times each century. Gentler tides that Jupiter causes in its moon Europa, and tides that Saturn produces in its moon Enceladus, sustain deep underground oceans of liquid water on those moons that continually shape their icy surfaces.

Like electrostatic force or the propagation of light, gravity as a force follows an inverse square law with distance. If, for example, you move 10 times further away from an object, you'll feel only one-hundredth the gravity you felt before moving. The mass of an object, on the other hand, is directly proportional to the gravity it exerts.

BIOGRAPHY

The German astronomer and mathematician **Johannes Kepler** (1571–1630) analysed decades of astronomical data in order to figure out the three laws of orbital motion that are named in his honour. The first and perhaps most fundamental law – that objects orbit their primary celestial bodies (the Moon's primary is Earth; Earth's primary is the Sun) in elliptical paths with the primary at one focus – is true because gravity follows an inverse square law.

The English astronomer and physical scientist **Edmund Halley** (1656–1742) made sure that the discoveries about the principles of physics and the mathematical explanations of motion and gravity discovered by Isaac Newton were published for the world to read. Halley then helped confirm Newton's theories by using them to calculate the orbital path of a well-known comet. He correctly predicted when it would reappear in Earth's night sky ; it has been called Halley's Comet ever since.

1,300,000,000 *years ago*

General Theory of Relativity

Isaac Newton explained gravity as a pulling force between massive objects. He was mostly correct.

Newton's model of gravity served well as a foundational theory of how things work in the universe for more than two centuries. For all the familiar physical systems in the cosmos – Earth, the solar system, the Milky Way and other galaxies – Newton's equations can be accurately used to predict just about all the motion we can observe.

About a hundred years ago, though, astronomers realized that Newton's law of gravity was a very good approximation but not exactly correct. Gravity, it turns out, is the consequence of space and time being closely tied to one another – an idea worked out by Albert Einstein called the general theory of relativity. The presence of a massive object, explained Einstein, puts a physical stress on space-time, in much the same way that a bowling ball might stress a trampoline. Just as that trampoline would curve inwards towards the ball, the resulting curvature of space-time inwards towards the object creates a 'dimple' into which nearby objects will fall – and the rate at which those falling objects will speed up matches exactly how they would move if they were being pulled by a force.

Beyond the general theory of relativity, the nature of gravity at the tiniest subatomic scales remains a frontier of theoretical physics. To this day, scientists are still trying to figure out how quantum mechanics and general relativity connect as two of the great contemporary theories of the universe. Quantum gravity may hold clues about why the Big Bang actually occurred, why space-time exists and if there are multiple universes and multiple dimensions of reality.

Light from a distant star can travel a bent path (red line) because space is curved by the mass of an intervening object. As a result, the star appears to be elsewhere (blue line).

Although the difference between Newtonian gravity and general relativity is almost indistinguishable, it clearly appears in the universe and in our daily lives. The clocks on GPS (Global Positioning System) satellites, for example, must be adjusted about 40-billionths of a second each day in order to measure locations on the ground accurately, because they experience the curvature of space-time just a little differently compared to objects at the surface of the Earth.

BIOGRAPHY

The English astrophysicist **Arthur Stanley Eddington** (1882–1944) strongly advocated for the general theory of relativity to be tested. In 1919, he led an expedition to measure the effect of the Sun's gravity on the perceived positions of stars in the sky visible during a total solar eclipse. When he and the Astronomer Royal Frank Dyson (1868–1939) announced their results confirming the general theory of relativity, it revolutionized the scientific world and made Albert Einstein an international celebrity.

The German-American physicist **Albert Einstein** (1879–1955) contributed many of the best-known advances to theoretical physics in the twentieth century. In addition to explaining the photoelectric effect, Brownian motion and the famous equation $E=mc^2$, he published in 1905 what is known today as the special theory of relativity. In 1915, he published the complex set of mathematical relationships describing the interconnectedness of space, time and gravity that are now called Einstein's field equations – the basis of the general theory of relativity.

Binary Stars

Most of the more massive stars in the universe are members of binary or multiple systems, where two or more stars orbit one another over long periods of time.

According to current estimates, about half of the Sun-like stars in the Milky Way also have close companions; for instance, Alpha Centauri, the closest known star to the Sun, is a multiple star system with three known stellar components and at least two planets.

How do binary star systems form? Over billions of years, stars moving around in galaxies will encounter one another as they orbit. Almost always, the stars are too far apart or moving too fast to do more than pass by, perhaps changing their respective paths slightly with their gravitational influence. When more than two stars interact at once, however, a complex interchange of momentum and kinetic energy can occur where two stars settle into orbit around one another and the other stars are flung away far and fast. (Conversely, a binary system can also be disrupted when a third star flies by, pulling the stars out of their mutual orbit.)

Most binary and multiple star systems are detached – that is, the objects in the systems don't touch. When the stars come close enough together, though, they can become contact binaries – like two balls of cotton overlapping to form an elongated blob – and matter from one companion may start flowing on to other one. And if one of the two stars is a compact object such a white dwarf, neutron star or black hole, matter will stream on to that object, releasing more gravitational potential energy than even the nuclear fusion inside the larger star. The result is a spectacular X-ray binary, which can have all the characteristics of a quasar at a millionth the size and power: a blazingly hot accretion disc, powerful jets and hard radiation emitted by multimillion-degree interstellar gas.

BIOGRAPHY

Italian-American astronomer **Riccardo Giacconi** (1931–2018) pioneered the use of rockets and satellites to conduct astronomical research, in particular the field of X-ray astronomy, which cannot be done within Earth's atmosphere. His experiments led to the discovery of many new types of objects such as X-ray binary star systems, including black holes such as Cygnus X-1. In 2002, he was awarded the Nobel Prize in Physics.

Does our Sun have a binary companion? If it does, it is too faint and too distant to be detected using current technology. Although there is no scientific evidence to support it yet, some have guessed there may be a very dim dwarf orbiting thousands of times further away from Earth than any of the known planets in the solar system. If it does exist, each orbit of that star around the Sun would take millions of years.

Cygnus X-1, a binary star system 6,100 light years away from Earth (artist's impression pictured opposite), was the first likely black hole candidate found in the Milky Way galaxy. Observational data show two objects, each more than 20 times the mass of the Sun, orbiting one another every five and a half days; one object is a bright blue star, while the other object is not visible and calculated to be at most the size of Ireland. Cygnus X-1 emits vast amounts of X-rays (hence its name), which are probably produced in and around the accretion disc surrounding the binary component that is a black hole.

Binary stars can theoretically also form together, perhaps in an extra-large cloud of interstellar gas. Such birth pairs don't have to be the same size, and if they merge together later in life it may help foster the creation of very high-mass stars that would otherwise be hard to form within their very short main-sequence lifetimes.

900,000,000 *years ago*

Black Hole Collisions

Stellar binary systems that feature two black holes orbiting one another are likely the rarest of the rare.

Astronomers wondered for decades if the physical environments of galaxies like the Milky Way would be able to produce even one such pair in its entire 10 billion-year lifetime. Nevertheless, the idea of black hole binaries fascinated scientists because if they did exist, then that would mean black holes could collide, creating the most energetic events in the cosmos.

The possibility of detecting such a collision was so enticing that a vast international team of scientists and engineers committed to creating an experimental facility to detect such a black hole collision, no matter how remote. Carefully calculating through Einstein's field equations of general relativity, they figured out that during the last fractions of a second before two black holes make contact, their inward spiral would release a pattern of gravitational waves that could be measured. If the pattern were converted into sound, it would even make an audible 'chirp'. The challenge was to build a set of ears, attuned to the cosmos, big enough and sensitive enough to hear it.

After years of effort and planning, construction of the Laser Interferometer Gravitational-wave Observatory (LIGO) broke ground in the 1990s. Upon its completion, scientists listened for gravitational waves in vain for more than a decade as LIGO continued to be improved and modified. Finally, on 14 September 2015 they heard it: in a distant galaxy, two black holes in a binary system, each about 30 times the mass of the Sun, had spiralled together into one, and LIGO had detected the ripples that collision had sent through space and time. The event was named GW150914, derived from 'Gravitational Wave' and the date of its observation (2015-09-14).

In the years since, dozens more gravitational wave chirps have been detected with LIGO and other gravitational wave observatories worldwide. These cosmic chimes may take decades to analyse and interpret, and represent an entirely new way to study the universe.

Black holes are concentrations of matter so dense that not even light can escape from their event horizons. The gravitational 'dimples' they make in space-time are small but deep; and if two deep dimples overlap, gravitational waves begin to radiate from the system, sapping the orbital energy of the binary and bringing them even closer together. Imagine two heavy cannonballs separated on a large sheet of fabric; if they come together, their two separate dimples come together, too, and waves of energy would pulse outwards, sailing through the fabric.

LIGO has two huge listening posts in the United States – one in Hanford, Washington, and the other in Livingston, Louisiana – separated by more than 3,000 kilometres (1,800 miles). Each observatory station has super-sensitive, laser-aligned 'eardrums' in underground vacuum chambers 4 kilometres (2½ miles) long. Each station is able to detect a gravitational pulse less than one hundred-millionth the diameter of a hydrogen atom. When a gravitational wave from a distant black hole collision passes through our planet, the two stations act in concert the way a pair of human ears would, comparing the signals they receive to decipher the location and power of the event they just detected.

BIOGRAPHY

Scottish physicist **Ronald Drever** (1931–2017) and American physicists **Barry Barish** (b. 1936), **Kip Thorne** (b. 1940) and **Rainer Weiss** (b. 1932) devoted decades of their careers to the development, construction and operation of LIGO. Their efforts were rewarded when the first gravitational wave detection occurred on 14 September 2015, followed by a number of additional detections in the following months. Just a few months after Drever died at the age of 85, Barish, Thorne and Weiss were awarded the Nobel Prize in Physics for their monumental work.

Two black holes collide

Gravitational waves emanate outwards from the collision

Far away, gravitational waves distort space
around Earth and are detected by LIGO

700,000,000 *years ago*

Meanwhile on Earth...

Oblivious to the amazing black hole collision that had occurred, Earth and the carbon-based life on its surface continued to experience its own unfolding story.

When the collision happened some 1.3 billion years ago, our planet was in the middle of what scientists jokingly call the 'Boring Billion' – roughly 1,000 million years of relative stasis, when both geological and biological changes were very slow and gradual.

Not long after that period finished, roughly 600 million years ago, paleontologists think multicellular life took hold on Earth. Around that same time, geological evidence suggests that Earth's climate grew colder, perhaps due to a dip in the amount of carbon dioxide entering the atmosphere from volcanoes. The resultant 'Snowball Earth' stage may have spurred a substantial amount of evolution by natural selection, and once the planet warmed up again a vast variety of life burst forth – a period 540 million years ago called the Cambrian Explosion.

Just about all the main categories of carbon-based life that exist on Earth today can trace their origins to the Cambrian Explosion. It was hardly smooth sailing for life after then, though. There were at least five periods of history, each lasting thousands or even millions of years, when most species of living things died out. Still, life as a whole persisted after these mass extinctions, and evolutionary milestones continued to appear from time to time. Land animals and plants proliferated around 480 million years ago; wood first appeared about 400 million years ago; leaves, around 360 million years ago; and the first flowers began to bloom about 130 million years ago.

Earth itself, meanwhile, kept evolving, too. Over the aeons, tectonic activity moved the vast solid plates of Earth's crust back and forth, similar to the way ice chips might float around on the surface of a punchbowl. When the first detected binary black hole collision occurred 1.3 billion years ago, a huge supercontinent called Rodinia was forming; by 670 million years ago, it had largely broken up. Its successor, Pannotia, broke up around the time of the Cambrian Explosion. The last supercontinent that concentrated all the world's major landmasses into one unit was Pangaea, which slowly drifted into pieces after the Permian Extinction, leaving the continents we recognize today.

Fossils provide a valuable but uncertain picture of life in Earth's distant past. Unlike redshifted light from distant galaxies and quasars, which is well-preserved radiation that has travelled mostly through the vacuum of space, fossils are usually greatly altered from the organisms' original composition. Many softer or more fragile parts of living things cannot survive the fossilization process and are thus lost to prehistory.

During the Permian Extinction, which happened about 252 million years ago, more than 95 per cent of all species of life in the oceans and two-thirds of species on land were extinguished. Geological evidence shows that vast volcanic activity occurred then, spewing enough lava to pave over at least 3 million square kilometres (1 million square miles). The fierce volcanic heat then ignited huge stores of fossilized plant and animal life that had been transformed into coal, oil and natural gas – the first-ever global burn of fossil fuels. The resulting ecological disaster, including extreme climate change and ocean acidification, dealt a serious setback to life on Earth, which took millions of years to recover.

A scientific reconstruction of Earth as it may have appeared between 500 million and 600 million years ago. Over eons of time, our planet's continents and oceans have moved dramatically, albeit very slowly.

67,000,000 *years ago*

66 million *years ago*

10

170,000 Years Ago

Sanduleak −69 202 becomes Supernova 1987A

A blue supergiant explodes, creating the first supernova in the history of modern astronomy visible to unaided human eyes.

As the gravitational wave pulse that became GW150914 continued at the speed of light in its billion-plus-year journey towards Earth, our planet – and solar system and galaxy – continued to age and evolve on their own timescales. The Sun travelled in its orbit around the centre of the Milky Way, making one full circuit every 250 million years, carrying its system of planets, asteroids and comets around for the ride.

All along its path, the Sun passed other stars in varying stages of life. Most were fainter than it, long-lived and stable, but others were massive, bright and volatile. Still others had already finished their stellar evolution and reached their endpoints. Intermediate-mass objects were now white dwarfs; high-mass stars had become neutron stars; and the highest-mass stars had evolved into black holes.

Somewhat unlike a dead carbon-based life form, though, a stellar remnant is far from an inert, decaying body. White dwarfs, neutron stars and black holes are comprised of matter in states too exotic to exist on Earth, and they continue to affect their environments actively – sometimes even more powerfully than their progenitor stars did when they fused hydrogen to produce energy. Some of them spin at breathtaking speeds; others harbour enormously strong magnetic fields and withering jets, and can even create matter from the energy they generate.

It may thus come as no surprise that the most powerful events that Earth has witnessed within its own galactic environment have been the explosion of stars – both dead and living. At the intersection of quantum mechanics and relativity, a special limit marks the boundary between what can remain a white dwarf and what must turn into a neutron star; at that boundary, stellar cores and corpses alike will detonate, creating a supernova so energetic that for a few seconds it outshines all the other stars in the Milky Way combined. Young neutron stars formed this way can pulse energy in lighthouse-like beams as they rotate – and, in very rare instances, form binaries that can simultaneously generate both gravitational waves detectable by LIGO and radiation visible to telescopes.

Fortunately, none of these titanic events we know of have occurred close enough to Earth to damage life here. That doesn't mean that terrestrial life hasn't ever been endangered by cosmic events, though. Sixty-six million years ago, a wandering solar system object crashed into us. Today we call it the Chicxulub impact, and it catalysed a mass extinction event that wiped out three-quarters of all the living species on the planet.

Millions of years later, one mammalian species that did survive would evolve into hominids – ape-like creatures who were our distant genetic ancestors. Then about 170,000 years ago, as the GW150914 pulse began passing through the Magellanic Clouds, a star in the larger of those two Milky Way satellite galaxies went supernova. Would that cataclysm lead to life's destruction? Happily, no – and even more happily, when its light reached Earth on 23 February 1987, we humans had evolved and prepared enough to recognize what it was and to study it with our own eyes, both unaided as well as augmented by modern scientific technology.

In 1054 CE, a new star appeared in the sky towards the constellation Taurus. Centuries later, astronomers discovered it was a supernova that had left behind this beautiful gaseous remnant: the Crab Nebula.

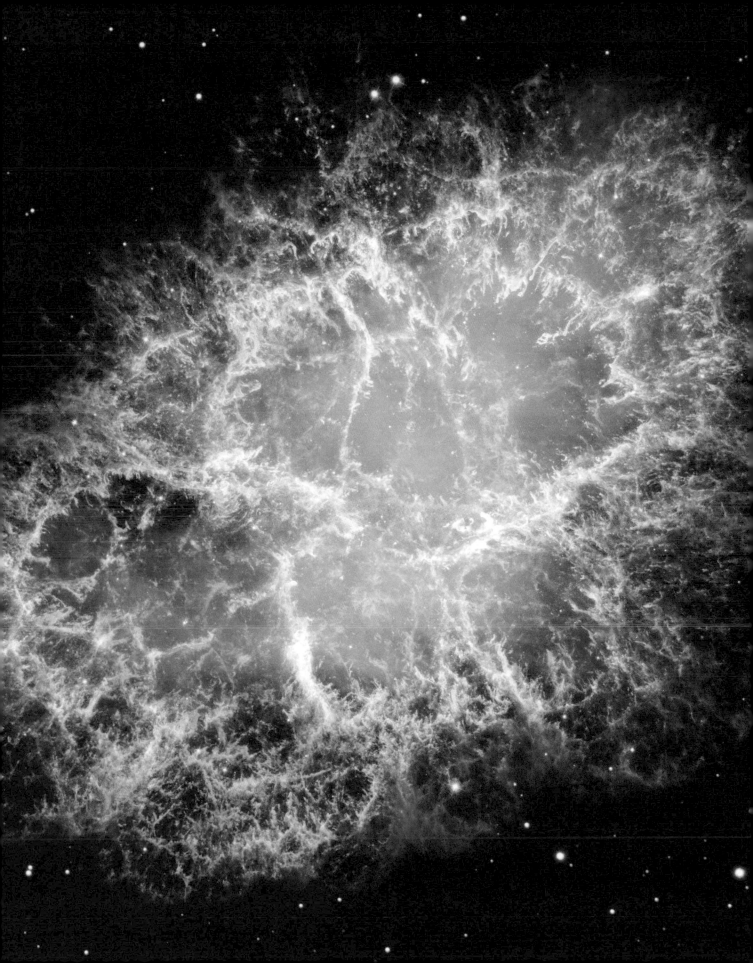

Dinosaurs Die and Mammals Rise

As multicellular carbon-based life continued to develop on Earth, plants and animals grew in complexity and size.

The largest animal life forms that ever developed on land were dinosaurs, some of which grew to more than 30 metres (100 feet) in length and weighed more than 100 tons. Alas, a mass extinction event about 66 million years ago called the K–Pg (Cretaceous–Paleogene) or K–T (Cretaceous–Tertiary) Extinction wiped out 75 per cent of all the species on Earth, including every large dinosaur species. The surviving dinosaurs evolved into the birds of today; and evolutionary biologists think that the surviving mammals, which at that time were generally small and unremarkable creatures, evolved into new large animals such as whales and elephants.

Around 55 million years ago, the earliest ancestor of one kind of mammal called primates began to walk the Earth. From humble origins, living in forests and barely the size of mice, they evolved a branch of themselves called simians; 25 million years ago, a branch of simians evolved known as apes. Eighteen million years ago, the hominids evolved, splitting further evolutionarily into orangutans (15 million years ago), gorillas (7 million years ago) and chimpanzees (3 million years ago). From then until now, all the remaining species of hominids have died out save only one: *Homo sapiens sapiens* – humans.

Direct evidence of a titanic meteorite impact was discovered by geologists who found huge ringlike features in the bedrock near Chicxulub, on the coast of southern Mexico, that were some 160 kilometres (100 miles) across and 30 kilometres (12 miles) deep. In 2016, a drilling expedition obtained samples from deep below the ocean floor, and found convincing evidence that an object at least the size of Mount Everest, falling from space at many thousands of kilometres per hour, struck that part of the Earth about 66 million years ago, producing a huge crater.

Important evidence that a meteorite strike caused the K–T Extinction is a particular layer of rock, in many places just a couple of centimetres or so thick, spread all over the world. Geological evidence shows that this layer was produced about 66 million years ago, and contains such a high concentration of the metal iridium that it was most likely ordinary rock that launched into Earth's atmosphere from the Chicxulub impact, mixed with the metal that had been in the vaporized meteorite, and then fell back to Earth over many years all over the globe.

BIOGRAPHY

American physicist **Luis Walter Alvarez** (1911–1988) was a pioneering experimental nuclear physicist. He was awarded the 1968 Nobel Prize in Physics for the development of the liquid hydrogen bubble chamber, a way to view the products of subatomic particle collisions. In 1980 he published, together with geologist **Walter Alvarez** (his son; b. 1940) and American nuclear chemists **Frank Asaro** (1927–2014) and **Helen Michel** (b. 1932), a paper describing how the extinction of the dinosaurs could have been caused by the collision of a meteorite about 10 kilometres (6 miles) across.

Stellar Deaths, Remnant Lives

During their lives, stars fuse hydrogen into helium, producing the energy that pushes outwards and holds the star in equilibrium against gravity.

The helium, though, can't really go anywhere, and over time collects in the centre of the stars where they were formed. If a star is massive enough, it will undergo a red giant phase when the helium can eventually be used as nuclear fuel as well, fusing to form carbon, which can then further be transformed into nitrogen and oxygen. If it's not more than about eight times the mass of the Sun, that's all the fusion that can happen with the temperature and pressure at the stellar core; as the star dies, the outer layers of the star blow away, leaving the helium or carbon and oxygen with only a tiny bit of hydrogen on top.

Unable to support its own weight, gravity crushes the stellar remnant until it becomes so dense that a teaspoonful of it would weigh thousands of kilograms. Left unchecked it would collapse to a single point. At this threshold, however, a quantum mechanical process takes over: the atoms are so tightly packed together that the electrons in the star begin to push against one another. Astronomers call this phenomenon 'electron degeneracy', and it produces an outward pressure that can balance gravity's inward force.

The end result is a blazing hot sphere barely the size of Earth – less than one-hundredth the star's original diameter. A 'white dwarf', as it is called, can start out with a surface temperature of 100,000°C (180,000°F) or more, and after hundreds of millions of years of cooling it can still be hotter than the surface of the Sun.

On average, white dwarfs are a little more than half the mass of the Sun; however, the nearest known white dwarf is Sirius B (pictured opposite, at the correct size relative to the Sun), a little less than nine light years away from Earth, and it is almost exactly one full solar mass. As its name suggests, Sirius B is a companion to the star Sirius – the brightest star in the night sky as seen from Earth. Indeed, a large fraction of known white dwarfs are in binary systems with other stars; and if the white dwarf's companion is still fusing hydrogen, and the two objects are close together, some of the star's gas may flow on to the white dwarf, slowly increasing the dwarf's mass. These binary systems are called cataclysmic variables, and in some cases, the newly acquired hydrogen superheats and produces bursts of light and even explosions of thermonuclear fusion at the white dwarf's surface. Such a flash is known as a nova – a Latin word meaning 'new' and a fitting description for fresh activity from a formerly dead star.

One oddity of white dwarfs is that, due to the way electron degeneracy works, the more massive they are, the smaller they are. Our own Sun will become a white dwarf someday, about 5 billion years from now.

SS Cygni is a famous cataclysmic variable – an average-mass white dwarf being orbited by a low-mass main-sequence star at a distance only half that between Earth and the Moon. Discovered by the American astronomer Louisa D. Wells in 1896, it is a 'dwarf nova' system that has been producing outbursts every few months for more than a century.

BIOGRAPHY

Austrian physicist **Wolfgang Pauli** (1900–1958) was awarded the Nobel Prize in Physics in 1945 for his discovery that fermions – a category of subatomic particles that includes quarks and electrons – must maintain their own distinct quantum properties, no matter how many of them are packed together. The Pauli exclusion principle explains why electron degeneracy pressure occurs to keep white dwarfs from total collapse.

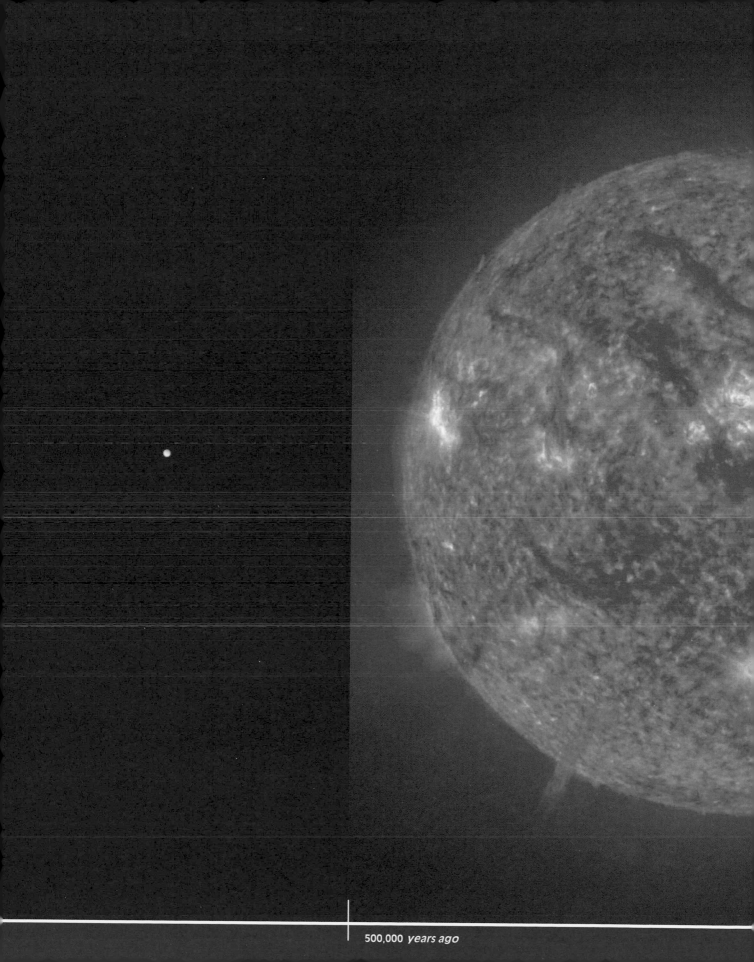

500,000 *years ago*

Type Ia Supernovae

The more massive a white dwarf is, the denser and hotter it gets.

If a white dwarf made primarily of carbon and oxygen reaches a density greater than 3 million times that of water as well as a temperature exceeding 500 million degrees Celsius, the carbon in the white dwarf can start to fuse together into heavier elements such as neon, sodium and magnesium. These thresholds are reached when white dwarfs reach a mass of about 1.4 times that of our Sun.

White dwarfs in binary systems can gain mass from their companion objects. If the companion is a large, mostly gaseous star, the white dwarf steadily accretes mass until it reaches the carbon fusion threshold. If the companion is a second white dwarf, on the other hand, the pair can spiral inwards and crash into one another, adding their masses together in a single event that could push the mass past the carbon fusion limit almost instantaneously.

Hydrogen fusion at the surface of a white dwarf creates a powerful but limited explosion, resulting in a nova. When carbon fusion begins in a white dwarf, however, a chain reaction ignites that causes fusion to run away, across and through the entire body of the dwarf. Within seconds, almost all of the carbon and oxygen in the white dwarf is consumed in a conflagration of fusion. The resulting energy release – more energy than our Sun will produce in its entire lifetime – blows the white dwarf apart. This is a supernova – specifically, a supernova of type '1-A', written 'SN Ia' – the ultimate exclamation point to a white dwarf's existence.

The nuclear fusion processes in type Ia supernovae produce almost all of the iron atoms that currently exist in the universe. Over billions of years, iron has been spread by these explosions throughout space, and has been incorporated into everything from the cores of Earth-like planets to the haemoglobin in human blood.

The critical mass for runaway carbon fusion is the same for any white dwarf. Type Ia supernovae are thus theorized to produce about the same amount of energy over the same amount of time and reach the same peak luminosity no matter where or when they occur. Astronomers have used this 'standard candle' property to make precise distance measurements to faraway galaxies; and the data have shown, to the surprise of many, that the expansion rate of the universe is speeding up. Type Ia supernovae have therefore provided key scientific evidence for the existence of cosmological dark energy.

BIOGRAPHY

The Danish astronomer **Tycho Brahe** (1546–1601) made years of detailed observations of the night sky that led to the discovery of the laws of orbital motion. In November 1572, he recorded the appearance of a star 'new and never before seen by anyone in living memory' and studied it extensively for more than a year. Modern astronomers now understand this event was a Type Ia supernova that occurred about 8,000 light years away from Earth, which left a fast-expanding gaseous nebula in the direction of the constellation Cassiopeia.

In October 1604, another Type Ia supernova detonated in the Milky Way galaxy – this one about 20,000 light years away. It has been called 'Kepler's Supernova' in honour of German astronomer **Johannes Kepler** (1571–1630), Tycho Brahe's protégé and successor, who published more than a full year's worth of detailed observations of the event.

A white dwarf accretes mass from its stellar companion

If it accretes enough material to reach critical mass,
the white dwarf will detonate, producing a Type Ia
supernova and blasting its companion away.

400,000 *years ago*

Special Relativity and Type II Supernovae

A white dwarf can support itself against its own weight because the electrons inside it produce electron degeneracy pressure.

The more closely packed its electrons get, the faster its electrons must move around in order to sustain this pressure. In 1935, a 25-year-old astrophysicist named Subrahmanyan Chandrasekhar proved that if those electrons started moving too fast, special effects of Einstein's theory of relativity would begin to take hold, resulting in the failure of the star to withstand its own gravity.

What would happen if a white dwarf in a binary system gained so much matter from its companion star that its mass reached Chandrasekhar's limit? Unable to support its own weight, a white dwarf would literally collapse to one-billionth its original volume in a fraction of a second. That generally doesn't happen, though, because the critical mass that ignites carbon fusion in a white dwarf is just a hair lighter than the Chandrasekhar mass; and once the runaway fusion starts, the white dwarf blows apart in a Type Ia supernova explosion.

There is, however, another kind of physical system where such a collapse can happen. At the hearts of stars heavier than about 10 times the mass of our Sun, carbon fusion can occur in their cores – but unlike in white dwarfs, the energy release is buried deep inside the massive star and does not run away. For hundreds more years – not thousands or millions, though – the carbon fusion continues; then for at most a few more months, heavier elements such as oxygen can fuse, and then silicon and sulphur fuse into iron for just a day or so.

When that last fusion process ends, gravity crushes the star's core so powerfully that the electrons there are all moving almost at light speed – and there is not nearly enough electron degeneracy pressure to stop the gravitational effect. The core collapses at a speed exceeding 60,000 kilometres per second (37,200 miles per second), from the size of our planet Earth to the size of the city of Paris. And the rebound of everything falling on to that core and bouncing back out results in another kind of massive stellar explosion: a Type II supernova.

A key point of the special theory of relativity is that within our flexible four-dimensional space-time, no object with mass can attain the speed of light – 299,792,458 metres per second, or just over 670 million miles per hour. As objects move faster and faster and approach that speed, it takes increasingly more energy to speed them up; simultaneously, the length intervals they experience get shorter, while time intervals grow longer. Human beings experience these relativistic length contractions and time dilations, but they are so small as to be unnoticeable. After a transcontinental flight, for example, you will have aged a few billionths of a second less than your friends on the ground – hardly enough to compensate for eating the airline food. For electrons in a white dwarf, though, the result is dramatic: if its mass reaches 1.44 times the mass of our Sun – we call that the Chandrasekhar mass – the relativistic effect on its electrons cause them to lose their ability to produce any additional degeneracy pressure.

Could a Type II supernova have caused mass extinction on Earth? Astronomers have calculated that if such a supernova detonated about 70 light years away, it could have stripped Earth of its stratospheric ozone layer for thousands of years, causing a significant but not total mass extinction. The Hangenberg Event, an extinction that happened 359 million years ago, seems to have matched this environmental disaster; no corresponding supernova remnant, however, has yet been identified.

BIOGRAPHY

The Indian-American astrophysicist **Subrahmanyan Chandrasekhar** (1910–1995) was born in Lahore in modern-day Pakistan. He published his first scientific paper at age 20, on the interior structure of stars, and five years later discovered the upper mass limit of white dwarfs, beyond which relativistic effects would cause a white dwarf to self-destruct. At the University of Chicago, where he spent most of his career, Chandrasekhar contributed substantially to almost every area of theoretical astrophysics. In 1983 he was awarded the Nobel Prize in Physics.

A high-mass star's inner layers collapse

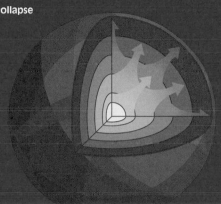

The stellar interior rebounds off the
collapsed core and blasts outwards

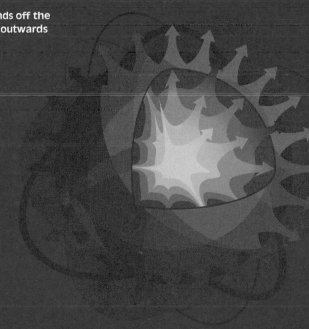

The star explodes creating a Type II supernova

300,000 *years ago*

Neutron Stars and Pulsars

Type II supernovae can generate at least as much energy as a Type Ia supernova.

In addition, the crushed, collapsed core of the Type II progenitor is left – a city-sized object containing more mass than our Sun. All atomic structure has been destroyed in this stellar remnant – all the electrons and protons are gone. Rather than electrons pushing against one another, neutrons are crowding together and providing neutron degeneracy pressure. This is a neutron star; and although it's tiny by stellar remnant standards, it really acts in many ways like one huge atomic nucleus.

One remarkable consequence of creating a neutron star in a Type II supernova is that any amount of spin that existed in the progenitor star is concentrated into the remnant. In the same way that ice skaters spin faster by pulling in their arms, the diameter of the stellar core shrinks so much and so fast that the neutron star can wind up spinning around and around, up to dozens or even hundreds of times each second. At such speeds, the spin generates a magnetic field that could be millions or billions of times stronger than Earth's field. The resulting electromagnetic dynamo is so energetic that photons can spontaneously transform into particles of matter and antimatter that cascade all around the neutron star and create bright spots that radiate radio waves, visible light and even X-rays and gamma rays. Every time the neutron star spins around, astronomers can observe a pulse of radiation coming from it.

On 4 July 1054, Chinese astronomers recorded a 'guest star' that stayed visible day and night for nearly a month. Nine centuries later, in 1968, American astronomers confirmed that the guest star was a Type II supernova. At the centre of the supernova's gaseous remnant, the Crab Nebula, a pulsar more massive than the Sun is spinning at 1,800 revolutions per minute – as fast as the engine of a running car.

BIOGRAPHY

The British astronomer **Jocelyn Bell Burnell** (b. 1943) was a graduate student at Cambridge University in 1967 when, examining the paper chart-recorders from radio telescope observations, she discovered a rapid, very regular pulsing signal coming from a spot in the sky in the direction of the constellation Vulpecula. The pulsing was so regular – almost exactly 1.3373 seconds apart – that she jokingly referred to it as a 'little green man', perhaps some sort of intelligent alien signal. She soon found other 'pulsars', as they became known, and they were confirmed years later as being rapidly rotating neutron stars. Later in her career, she served as President of both the Royal Astronomical Society and the Institute of Physics. In 2018, she was awarded the Special Breakthrough Prize in Fundamental Physics for her discovery of pulsars and her leadership in the scientific community.

A neutron star is as dense as an atomic nucleus. One teaspoonful of neutron star material would weigh some 5 billion tons – about the combined mass of every human being who has ever lived. If you poured that teaspoonful on to the ground, the little blob of material would cut through our planet as if it weren't even there; fall through the centre of the Earth and then come out the other side; stop, turn around, fall back through Earth and reach the height of your spoon; and then keep going back and forth unstoppably for thousands of years.

On 17 August 2017, the Laser Interferometry Gravitational-Wave Observatory (LIGO) heard a gravitational wave chirp; moments later, two telescopes in orbit detected a burst of gamma rays. Over the next several weeks, more than 70 observatories around the world and in space found and observed the source of this signal: the first gravitational wave pulse ever to have been detected with other telescopes, caused by the collision of two neutron stars 130 million light years away from Earth (artist's impression pictured opposite).

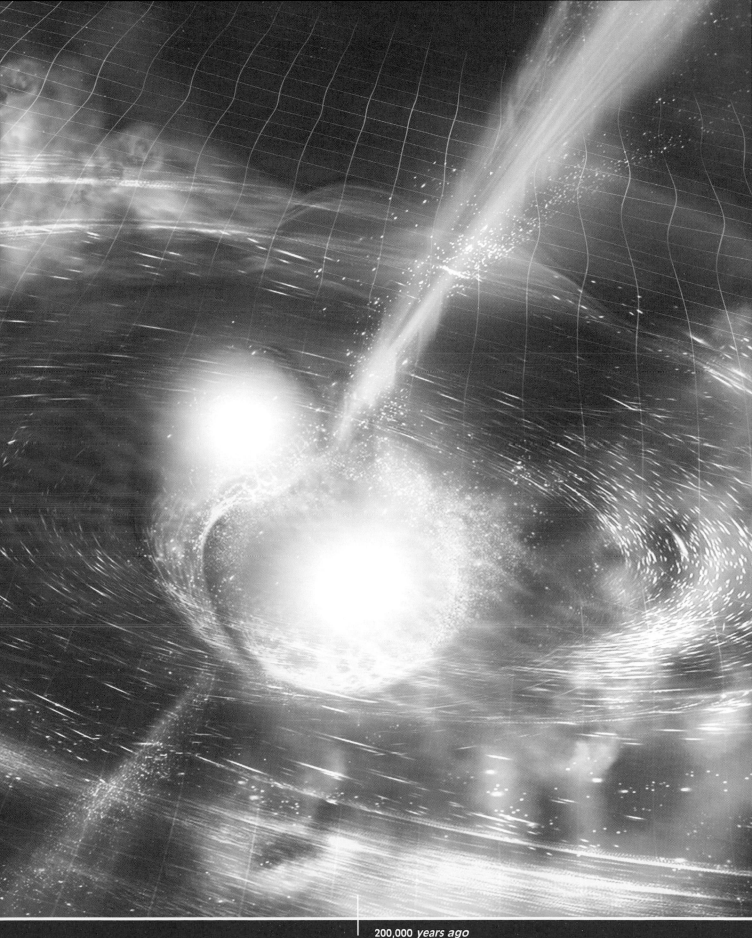

200,000 *years ago*

Magellanic Clouds and SN 1987A

Among the dozens of satellite galaxies that orbit our Milky Way, the two most prominent are beautifully visible with the unaided eye from Earth's southern hemisphere.

Named after Ferdinand Magellan (c.1480–1521), who led the first European circumnavigation of the world, the Large Magellanic Cloud is about 160,000 light years away from Earth and 14,000 light years across; and the Small Magellanic Cloud is 200,000 light years away and about 7,000 light years in diameter. Both galaxies have fewer than one-hundredth the number of stars as the Milky Way; but both galaxies are also rich with cold hydrogen gas – the raw material for starbirth – and thus contain large numbers of young, high-mass stars.

A little less than 170,000 years ago, one of those high-mass stars on the far side of the Large Magellanic Cloud – a bright blue star called Sanduleak −69 202 – exploded as a Type II supernova. When its light reached Earth on 23 February 1987, sharp-eyed astronomers in Chile and New Zealand spied it almost immediately – the first supernova visible without telescopes to be seen by humans in nearly four centuries. It was given the name SN 1987A, and by May of that year it had become about as bright as the stars in the Big Dipper.

Although SN 1987A has faded in brightness over the decades, it continues to be one of history's greatest-ever treasure troves of stellar scientific knowledge. Astronomers have been able to trace the ageing and evolution of the supernova since its first day, using modern telescopes, electronic cameras and space observatories; from those measurements, we have learned enough to literally rewrite the book about the life cycles of massive stars and how they meet their end.

Hours before the light from SN 1987A (circled in the photo) was first seen by humans, other astronomical 'eyes' had already detected the supernova explosion. About two dozen neutrinos mysteriously struck three different observatories – one in Japan, one in America and one in Russia – within seconds of one another. That was about 1 million times the normal detection rate of neutrinos from space, and the first time neutrinos had been detected from any cosmic source other than the Sun. Those detections showed that neutrinos streaming from the supernova explosion carried a hundred times more energy away from the collapsing stellar core than the heat and light that was emitted.

BIOGRAPHY

Working at the Harvard College Observatory, the American astronomer **Henrietta Swan Leavitt** (1868–1921) identified and studied more than 1,700 stars in the Small and Large Magellanic Clouds that varied in brightness. She discovered that a particular kind of variable star called Cepheids followed a mathematical relationship between their peak brightness and the time interval between their peaks. Today, the Leavitt Law that describes the Cepheid period-luminosity relation is a key to measuring distances to the galaxies beyond our Milky Way, and our distance to the Magellanic Clouds is the most important rung in the cosmic distance ladder.

The Romanian-American astronomer **Nicholas Sanduleak** (1933–1990) specialized in the spectroscopic study of stars. Among his work was a catalogue of interesting stars in the Magellanic Clouds; one of those stars, known today as Sanduleak −69 202, was a bright blue supergiant star that became the Type II supernova SN 1987A. It was the first star in scientific history for which astronomers had data before it exploded as a supernova.

100,000 *years ago*

90,000 *years ago*

11

20 July 1969

Apollo 11 – Humans Walk on the Moon

'Houston, Tranquility Base here. The Eagle has landed.'

Minutes after Neil Armstrong and Buzz Aldrin transmitted that message to Earth from the surface of the Moon, they took their very first steps on our celestial neighbour on behalf of all of humanity.

We have learned about cosmic evolution, galaxy evolution, stellar evolution and planetary evolution. What about human evolution? Of course, as a single species of carbon-based life on one small planet, whose individuals have a lifespan far shorter than even the shortest-lived star, we may seem to be too ephemeral and insignificant to consider. And yet, we are both as amazing and as ordinary as any star or galaxy, made of the same kinds of particles and affected by the same kinds of physical processes as anything else in the universe. So the study of our place in cosmic history, viewed through the lens of science, is indeed a worthwhile study of a piece of the universe itself.

Just as successive generations of stars differ because of the influence of their environments and their progenitors, modern humans are different from our early biological ancestors, shaped by our genetics and through evolution by natural selection. And just as with stars, human evolution is not about improvement but rather about change. Environmental conditions threatened our lives, and we made adaptations to survive. Our physical structures – bones, blood, brains and more – have all changed over

time. So, too, have our interactions with our fellow humans, as our communities and societies have morphed over the millennia. We have learned to use language to communicate, tools to construct, and science and technology to create and explore.

What has been the result so far? Today, for reasons we only barely understand, the evolution of our species has led us to a fascinating point in history. Unlike any other life forms on Earth, we have consciously and voluntarily decided to travel beyond the confines of the planet of our birth. Using our creativity and skills, we have built tools and machines to turn imagination into reality. We have sent a few of us to a world not our own, and brought them safely home again. And if the history of humanity is any guide, we won't be stopping there.

If *Homo sapiens* continues to survive and thrive, and we wind up exploring far beyond our world and even our solar system, one mystery will be at front of mind wherever we go. Out there, in the vast universe, will we ever find others like us? For all of our searching so far, the answer has been clear: no. We have yet to look, though, in so many places where we think extraterrestrial life could be. With our scientific tools and our active minds, we will continue the search; and until the answer is yes, we will probably never stop.

Astronaut Buzz Aldrin deploying scientific research equipment on the Moon near the lunar module *Eagle*, 20 July 1969.

Human Evolution and Migration

Modern humans can all trace their genetic heritage to a point approximately 3 million years ago, when the hominids split into two evolutionary branches.

One branch, the genus *Pan*, evolved to become chimpanzees. The other, the genus *Australopithecus*, evolved further into the genus *Homo*, which spawned a number of species that lived in communities and were able to use stone tools. The most successful of those species was *Homo erectus* ('standing human'), which appeared about 2 million years ago; they spread throughout Africa, their continent of origin, and then migrated to Asia and Europe over the next million or more years.

The exact dates of evolutionary changes in early human species are a frontier of modern scientific research, and the specific periods or events that led to rapid human evolution are still uncertain. What geological evidence shows is that around

780,000 years ago, Earth's magnetic field reversed its polarity, marking the start of the Chibanian (also called the Middle Pleistocene) age of Earth's history. By that time, early humans were using fire, hunting and gathering in groups, and caring for ill and injured members of their communities.

By the end of the Chibanian period, a wide variety of subspecies of *Homo erectus* had evolved, including 'Java Man', 'Peking Man', 'Rhodesia Man' and 'Heidelberg Man'. Some of these subspecies survived for hundreds of millennia. The last ones known to science to have died out included the Floresiens, the Denisovans and the Neanderthals. All humans today are of the subspecies called *Homo sapiens* – 'humans who can be wise'.

BIOGRAPHY

Lucy (reconstruction shown above) also called AL 288-1 (c. 3,200,000 BCE) was a female member of the *Australopithecus afarensis* species of early humans that lived in modern-day Ethiopia. First found in 1974, about half of her fossilized skeleton has since been recovered; she walked on two legs, and stood just over 1 metre (3 feet) in height. **Turkana Boy** (KNM-WT 15000, c. 1,600,000 BCE) was a young male member of *Homo ergaster* whose nearly complete skeleton was discovered in 1984 by the Kenyan archaeologist **Kamoya Kimeu** (b. 1939).

Homo sapiens are migratory animals with a history of exploration. Our earliest ancestors were born in Africa; waves of movement over a million years or more led to communities of archaic humans being established all over the world. At different geological periods when seas were shallower or land bridges developed, humans were able to walk or even take short boat rides from Asia to Australia (more than 40,000 years ago) and also from Asia to the Americas (at least 14,000 years ago). To this day, humans continue to immigrate across the globe, and some are even living for long periods of time in space.

Genetic research shows that in the last 100,000 years or so, interbreeding between related early human species occurred frequently. Today, all human populations not from Africa have some genes from *Homo neanderthalis*, and it is estimated that roughly 5 per cent of today's human genome originated from archaic humans. The definition of a species is based generally on both anatomical and genetic similarities; so in the same way that people with different hair or skin colour are all equally human, so, too, are people that might, for example, have different numbers of arms, legs, fingers or toes.

Many human populations carry genetic markers that can be traced all the way back to humanity's African birthplace. The arrows on this map represent plausible human migration routes based on DNA evidence: the blue lines are from Y-chromosome markers (passed down from father to son) and the red are from mitochondrial DNA markers (passed down from mother to child). Members of one branch of *Homo sapiens* apparently left Africa around 100,000 years ago, and gradually spread worldwide.

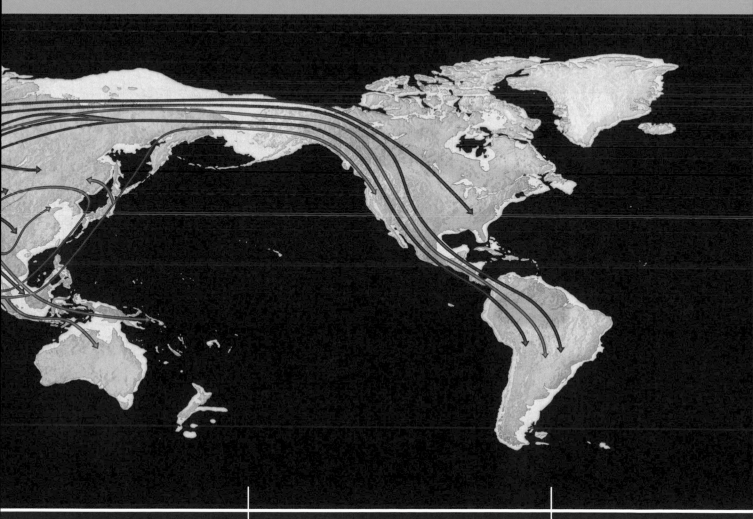

50,000 *years ago*   30,000 *years ago*

Society and Civilization

By about 20,000 years ago, the human species had established communities throughout the world.

Each early human community survived and reproduced by adapting to their environments and surroundings. In some parts of the world, it became advantageous for the humans there to find new ways to increase their ability to thrive by developing certain specialized activities, as well as the tools to perpetuate those activities for many generations.

Between 10,000 and 15,000 years ago, while most human communities hunted and gathered their food, a few groups began to domesticate animals to do work and ensure a more reliable food supply – pigs, sheep and cows in Asia; llamas, alpacas and guinea pigs in America. Around the same time, other groups began to grow edible plants such as wheat, peas, lentils and rice (and in America, potatoes). As humans developed more and more tools to help them do work, they also developed tools for play. An activity unusual among animal species began to occur with humans: artistic pursuits such as drawings and music appeared, which enhanced the communication of events and ideas and the connections between people.

Many scientists think that the development of complex spoken and written languages caused humans to be able to adapt and change extremely rapidly compared to other animal species. Figuring out when humans began to use language, however, has been an almost unsolvable scientific problem. (It's still very uncertain, for example, if animals other than humans use languages to communicate.) What we do know is that early writing had developed by about 8,000 years ago, and well-formed written languages were fully established in several human cultures as early as 5,000 years ago. This technology made it possible for humans to pass detailed records along to their descendants, so a sustained and growing body of knowledge could be widely distributed over many human lifetimes.

Over thousands of years, as human populations grew, communities further developed rules of behaviour that could increase the chances of survival of at least some of their members. Like other social creatures such as bonobos or bees, individual humans can learn specialized skills that help them serve useful roles within their societies. Among Earth's animal species, though, *Homo sapiens* seems to be uniquely flexible, with each person capable of learning and performing a wide variety of tasks and contributing to the well-being of their community in numerous ways.

BIOGRAPHY

The Belgian-Australian archaeologist **Andrée Rosenfeld** (1934–2008) was the daughter of two physicists, Drs. **Yvonne Cambresier** (1911–1988) and **Léon Rosenfeld** (1904–1974). Andrée earned a doctorate in geochemistry, and subsequently in her career applied her academic training to pioneer the scientific study, appreciation and preservation of prehistoric cave art and rock art, especially that which was produced by early Australians more than 10 millennia ago.

In 1799, a French military officer found a stone carved with Egyptian hieroglyphs, ancient Greek and ancient Egyptian Demotic script. What it said was a mystery for years, although the English physician and physicist **Thomas Young** (1773–1829) made significant progress translating it starting in 1813. The full translation of the Rosetta Stone was completed by the French linguist **Jean-François Champollion** (1790–1832), opening the door to the amazing history of ancient Egypt.

Around 3200 BCE, cuneiform, a written language finely carved into clay tablets (centre right, facing page), was developed in Mesopotamia, while hieroglyphics – which combined pictures and alphabetic symbols – arose in Egypt. The ability to read these languages was lost for centuries, and then regained in the nineteenth century by linguists who applied code-breaking methods to interpret the writing.

Science, Technology and Society

It's impossible to say when humans began to consider their own futures.

Beyond the basic activities common to all DNA-based living things such as survival and reproduction, the origin of the human development of technology and its use to plan far ahead remains a mystery. All we can say for sure is that it happened, and the application of scientific knowledge to produce new tools and materials made it possible. Around 5,000 years ago, some human communities learned how to produce temperatures hot enough to refine metallic ores such as copper and tin, and then combine them to make bronze – a hard and durable tool-making substance. Twenty centuries later, other groups began to refine and shape iron, an even harder metal. Over time, these and other materials were used more and more for artistic as well as utilitarian purposes. Communities expanded into societies, with thousands of humans doing specialized and interconnected tasks, and different groups of humans came to be expected to behave in specific ways.

By the time complex societies developed, with human populations numbering in the millions, it was becoming increasingly important to schedule and coordinate those tasks. This need led to the development of calendars, and the realization that the motions and patterns of celestial objects were excellent ways to mark the passage of time. In humanity's quest to understand and decipher the world and universe around it, the precursors of scientific disciplines began to develop; and eventually, astronomy replaced astrology and chemistry supplanted alchemy, while numerology and superstition were superseded by mathematics and physics.

The boy-pharaoh Tutankhamun (c. 1342–c. 1325 BCE) was mummified and buried in the Valley of the Kings, across the Nile River from the ancient Egyptian city of Thebes (modern-day Luxor). Buried with him was a dagger with an iron-nickel blade that was carved from a meteorite. Also among the antiquities in Tut's tomb was a lotus-shaped alabaster chalice with the inscription: 'May your spirit live, may you spend millions of years ... with your face to the north wind, your eyes beholding happiness.'

Human cultures worldwide used Earth's connection with the rest of the universe to keep time and create calendars. Structures such as Stonehenge in England, estimated to have been built between 3000 BCE and 2000 BCE, may have been used to mark the seasons and predict eclipses of the Sun and Moon. Many other ancient structures, such as the pyramids in Egypt and Central America or religious temples in Asia, were constructed with a keen awareness of the motions and positions of the Sun, Moon and stars (for example, Jantar Mantar in Jaipur, Rajasthan, India, bottom).

BIOGRAPHY

American anthropologist **Margaret Mead** (1901–1978) pioneered the study of the connections between human cultures and personal behaviours. While conducting field research in the South Pacific in the early twentieth century, she observed that the roles of men and women were not universal among all people, but instead naturally varied widely depending on the society in which they lived. Her advocacy of these and other conclusions raised great controversy in the Euro-American society of her time; nevertheless, she was posthumously recognized with the Presidential Medal of Freedom for her profound influence on social science.

Humans into Space and on to Other Worlds

By the second millennium CE, the human species was making a considerable effort to study and understand the universe in which it lived.

Major astronomical observatories were established during medieval times in places such as Dengfeng in China, Maragheh in Persia and Uraniborg in northern Europe. Then in 1610, the Italian astronomer and scientist Galileo Galilei (1564–1642) published *Siderius Nuncius* (The Starry Messenger) – detailing observations he had made of the heavens using a newly developed scientific instrument called a telescope.

Galileo's observations showed that the moons and planets of our solar system were not supernatural entities or unfathomable points of light, but rather entire worlds unto themselves. Not surprisingly, humanity began hoping and planning to visit those worlds. How, though, could we get all the way out there, much less come home safely again?

Slowly but surely, using science and technology, humans developed the tools to do just that. Gunpowder-based rockets ('fire arrows') were first developed by Chinese inventors more than 700 years ago. A few centuries later, visionary designers such as Leonardo da Vinci (1452–1519) and Conrad Haas (1509–1576) began to conceive of flying machines and liquid-fuelled rockets that humans could control to carry them as passengers into the sky. It would be a few more centuries – many human lifespans, yet just a blink of an eye in cosmic time – before those dreams were realized. In July 1969, three humans left Earth's embrace and went to the Moon; two of them walked on to the Moon's surface; and all three then came safely back home.

From 1969 to 1973, 12 people walked on the Moon. The human society that expended the energy and resources to send them there has since chosen to explore other worlds not with more people, but rather with remote-controlled machines that cost much less and pose no risk for the loss of human life. If society's priorities change, though, and enough people deem it worth the effort, then there will come a time when humans will once again stride on to the surface of another world.

BIOGRAPHY

The Russian teacher **Konstantin Tsiolkovsky** (1857–1935), French aviator **Robert Esnault-Pelterie** (1881–1957) and American physicist **Robert Goddard** (1882–1945) developed, independently and in quick succession, the mathematical formulation (known today as the 'ideal rocket equation') that shows how much speed a rocket can gain given its mass, the speed of its exhaust and the amount of fuel it's carrying. Goddard also designed and flew the world's first liquid-fuelled rockets, pioneering the technology that sent humans to the Moon four decades later.

On 17 December 1903, American inventors Orville Wright (1871–1948) and Wilbur Wright (1867–1912) made the first engine-powered, controlled atmospheric flights off the surface of Earth in their aeroplane *Flyer 1*. On 19 April 2021, NASA engineers conducted the first such flight off the surface of another planet with the *Ingenuity* Mars helicopter. Onboard *Ingenuity*, in commemoration of its historic legacy, was a small piece of fabric that covered part of the *Flyer*'s wing.

Premiere Montgolfiere, sans passagers
4 Juin 1735

UNITED STATES

USA USA

Are We Alone?

Throughout all our first scientific explorations into the universe, we have yet to encounter any other carbon-based life forms that did not originate on Earth.

We have therefore not yet succeeded in answering two of humanity's most omnipresent questions: is Earth the only place in the universe where there's life? And is our species the only one in the universe that is self-aware – that can communicate, that can imagine?

Our own cognitive limitations are surely one obstacle to our obtaining the answers we seek. We barely understand the definition of life for the carbon-based creatures on our own planet; given the myriad possibilities of what could have happened among all the worlds out there, would we even recognize extraterrestrial life if we saw it? Indeed, should stars and planets be considered living things? Perhaps, in the same way that trillions of micro-organisms have evolved to live inside each human's digestive tract, Earth itself may be alive – and we humans are simply too primitive to know how to talk to it.

Despite the challenge, humans have continued to work diligently to seek scientific evidence of life on other worlds. Starting in the 1960s, we sent *Venera* landers to the surface of Venus, and we sent *Viking* landers to the surface of Mars in 1976. In the following decades, we sent the *Galileo* probe to orbit Jupiter and its moons; the *Cassini* probe to Saturn and the *Huygens* lander to the surface of Saturn's largest moon, Titan; *Voyager II* to fly past the planets Uranus and Neptune; and even the *New Horizons* probe to fly past the dwarf planet Pluto and its five moons. We have aimed sensitive radio telescopes at nearby stars and exoplanets, listening intently for radio signals that might reveal the presence of creatures that process language or mathematics like humans.

So far, we have found no evidence to confirm the existence of extraterrestrial life. Undaunted, we continue to search with new tools and imaginative technologies. In the past few years, robotic rovers on Mars have found evidence that conditions on that planet could have supported living microbes there billions of years ago. Furthermore, astronomers know of thousands of planets outside our solar system, a few of which may be similar enough to Earth that they could harbour life as we know it.

As our species continues its search, there is reason to be optimistic that living things we could some day detect and contact do exist on other worlds. After all, the laws of nature work the same way everywhere in the universe – and we are here.

BIOGRAPHY

The American astronomer **Carl Sagan** (1934–1996) made important contributions to the Viking missions to Mars and to the scientific search for extraterrestrial life. In 1980, he hosted the popular science television show *Cosmos*, and became the world's foremost communicator of science to the public during his lifetime.

The American astronomer **Nancy Grace Roman** (1925–2018) conducted significant research on stars in the Milky Way similar to the Sun. In 1959, she published an article in *The Astronomical Journal* describing a method to detect 'planets of other suns' using a telescope outside Earth's atmosphere. A few years later, she helped create the space astronomy programme for NASA, serving as the agency's first Chief of Astronomy and laying the groundwork for today's fleet of orbiting space telescopes. NASA's next major space telescope, designed to take wide-field infrared images of the universe and to identify and study exoplanets, is named after her.

Searching for life in our solar system. Above, colourful streaks on Jupiter's moon Europa hint at the possibility of life under its icy crust; below, this mosaic 'selfie' of the *Perseverance* rover on Mars on 6 April 2021.

Tomorrow

12

The Future
of the Universe

From Now to the End of Cosmic Time

Thanks to astronomy, we can scientifically predict the likely future of our observable universe.

Thousands of years ago, before humans had developed the methods and tools of science and technology, we wondered about the unknown and tried to make sense of it with what facilities and resources we had. By observing the patterns of nature around us, on the Earth and in the sky, we soon gained the power to predict the future. When would be best to hunt and fish to get the most food? When should we plant crops to ensure the most abundant yields? When could we expect warm temperatures, or frigid conditions, or seasons of rain or drought?

Our ancestors' efforts to pin down the exact hours or days of specific events were crude, but over the years and centuries the rhythms of the universe gave them enough data to create civilization as we know it today. Cosmic processes seemed especially reliable; whereas knowing whether the weather tomorrow would be cloudy or clear was challenging at best, we were able to say precisely what time daylight would start, when it would end, how high in the sky the Sun would get at noon, and even when the Moon would rise and what it would look like. No matter what happened in the affairs of our lives, it seemed that the cosmos would stay steady and unchanging, ever aloof – and, perhaps, even supernatural or divine.

Today we know better. Certainly, so much remains unknown in the vastness of the universe that we would be rash to stop seeking new knowledge and simply accept things as unknowable. Just as scientific inquiry has always shown us, however, the process of discovery inevitably leads to new questions. This much we think we know: our small volume of the universe will stay roughly the same for millions of years. Then, a very long time from now, Earth's surface will be rendered inhospitable to life as we know it by a slowly brightening Sun. Then the Sun itself will die while, at about the same time, the Milky Way galaxy crashes into the Andromeda galaxy. Then, much later, the stars in the sky will all burn out. After that, we are less sure, but we think that all of the atoms and molecules in the universe will fall apart, and then every black hole will dissipate into nothingness.

How confident are we that cosmic history will unfold in this order? Ironically, the part of the future we can least accurately predict is the path directly in front of us. Just as our ancestors could predict the seasons but not the weather, we know what will befall Earth in millions of years but not in the next century. Blithely vaulting past our near-term uncertainty, we can look forward many trillions of years with reasonable confidence, at least in broad strokes. And then, as we head into the truly far future, our crystal ball turns hazy again, at least until more scientific studies are conducted and we can solve a few more mysteries of the universe: proton decay, Hawking radiation, dark matter and dark energy.

Maybe the best news about the future is that the universe will continue to expand as far into the future as we can currently calculate. That means, in our ongoing quest to understand the cosmos, we can afford to be persistent and patient – we have all the time in the world. The downside is that for some puzzles, that may still not be enough.

The Future of Humanity

Human evolution has brought our species to a fascinating point in cosmic history.

For the first time on Earth, an organism has the ability to leave its original ecosystem entirely – to go away from the planet and survive beyond its evolutionary confines. Simultaneously, that same life form can cause a mass extinction that would wipe out every member of its own species – not over eons or millennia, but within a few years or even a few days; and it could achieve those results either accidentally or intentionally.

Humans have demonstrated a remarkable characteristic that appears almost nowhere else in nature: individual members of our highly social species can behave in just about any way they choose, regardless of what other members are doing. Perhaps even more remarkably, that ability has not weakened humanity's prospects for survival, but rather enhanced them; creativity and imagination have combined with an uncanny ability to use tools and make machines to do and build what no single person alone could ever achieve.

Thanks to science and technology, humanity has cured diseases, increased the quality of life for most of us, and enjoyed its first taste of cosmic exploration. Sadly, the same science and technology can cause tremendous harm if used unwisely or inconsiderately – to ourselves and, intertwined as we are in the planet's complex ecosystem, even to all life on Earth. Our species has made it this far without going extinct; if we continue to learn about how our universe works, and then use that knowledge carefully while being considerate of our individual and collective responsibilities to the world and to one another, we can hope to keep our species alive and thriving for a long time to come.

Humans may already have enough knowledge to forestall one kind of mass extinction threat. The K–T Extinction, which happened about 66 million years ago, was triggered by the collision of an asteroid or comet about 10 kilometres (6 miles) across. With enough planning and appropriate resources, astronomers can detect a potential impactor of that size and possibly divert it harmlessly away from Earth. Whether the human species will collectively make those resources available in time to prevent the next such collision remains to be seen.

While protecting and improving our living environment on Earth remains far easier for the time being, looking outwards towards other places where humans could live might increase humanity's odds of surviving planetary catastrophe. Scientists estimate that with a century of careful and deliberate effort, humans may be able to make sustained living communities on the Moon or our neighbouring planet Mars – not as colonizers, but as respectful immigrants and explorers.

Humans have changed the world more in the past two centuries than any other species on our planet, primarily due to the prodigious use of fossil fuels. One waste product of burning those fuels, however, has been vast amounts of carbon dioxide (CO_2). By 1950, the amount of CO_2 in Earth's atmosphere reached a level not seen for at least 800,000 years; and in the past 70 years, the level has increased another 40 per cent. The extra insulation caused by that excess CO_2 means that there are about 100 million atomic bombs' worth of additional heat coursing through the air now compared to our grandparents' time.

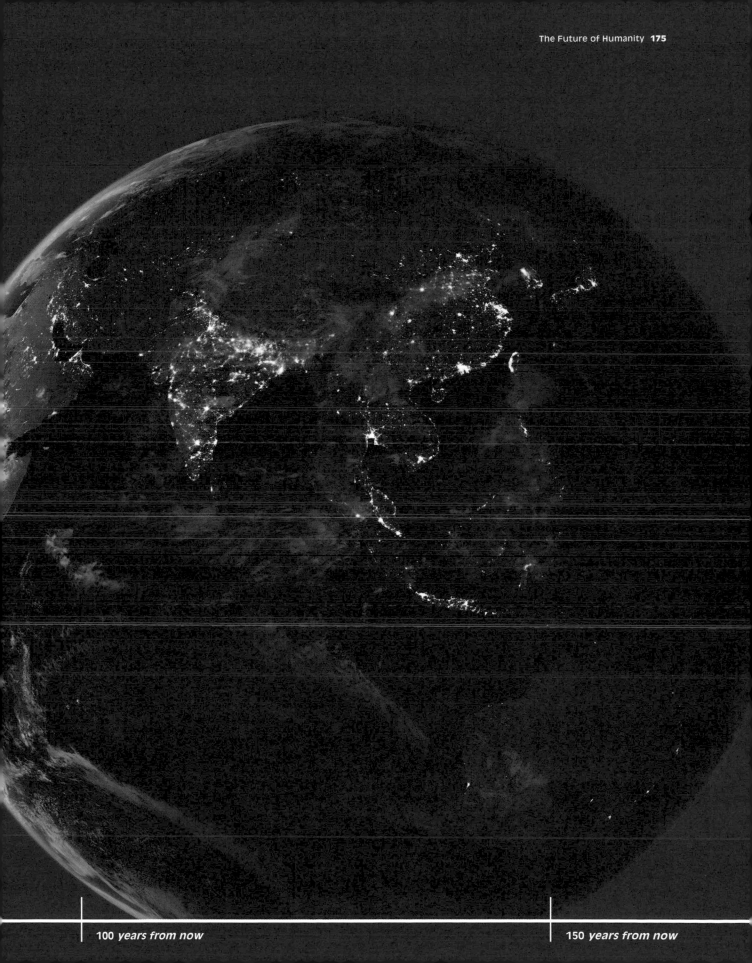

The End of the Sun

The Sun's constancy is remarkable by human standards.

In all of recorded history, the total amount of solar energy that has arrived at Earth each year has never varied by more than a fraction of a per cent. That is of course a good thing for our survival and general well-being; it has not always been this way, however, nor will it stay that way into the far future.

Ever since it began its nuclear fusion-powered life, the Sun has been getting brighter at the rate of about 0.006 per cent per million years. Fortunately for life on Earth, our planet in the past emanated much more warmth from within compared to the present day, as the heat of its formation escaped through the atmosphere and into space. The Sun's gradual heating thus balanced Earth's slow cooling, keeping conditions at Earth's surface reasonably steady. Before too long, alas, the equilibrium will be upset and Earth will get steadily hotter. At the current rate, Earth's oceans will be boiled away 1–2 billion years from now, and life as we know it will not be sustainable.

Three billion years after that scorched-Earth scenario, the very existence of our planet will be threatened by the Sun. Our host star's main-sequence lifetime will be finished, and as its interior processes adjust to that transition it will swell to a million times its current volume as it becomes a red giant. Its radius will grow one hundredfold as it engulfs and incinerates the planet Mercury, and it will blast Venus, Earth and Mars at close range with blistering heat and powerful solar winds. After a billion years of this extreme environment, any extant Earthly remnant will undergo a final assault as the Sun's outer layers are expelled outwards into a planetary nebula, exposing its hot white dwarf core.

Sunspots are huge magnetic storms in the Sun's outer layers that can last for weeks and often exceed the size of our planet Earth. Strong sunspot activity correlates with a higher flux of solar radiation, and when solar material is hurled towards Earth the result can be spectacular displays of the aurorae, the northern and southern lights.

The cataclysmic demise of our Sun and our planet need not result in a similar end for humanity. Although 1 billion years may be a brief interval of cosmic time, it is thousands of times longer than *Homo sapiens* has even existed as a biological species; indeed, 1 billion years ago today, the most evolved life form on our planet only had one cell. If humans are still around when Earth's surface is baked dry, a combination of biological evolution and technological advancement will by then almost certainly have enabled our distant descendants to escape the heat.

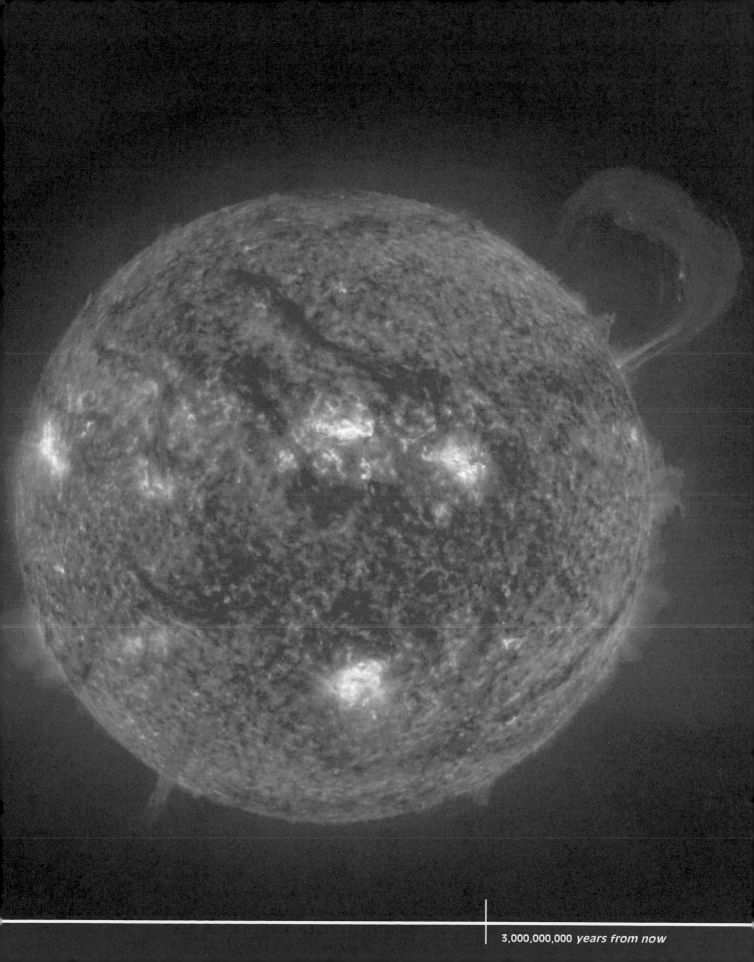

3,000,000,000 *years from now*

The Milky Way and Andromeda Collide

As the Sun approaches the end of its main-sequence lifetime, an even grander cosmic demise will begin all around our solar system.

The Milky Way galaxy, in which our solar system resides, has been steadily moving towards the constellation Andromeda at more than a quarter of a million kilometres per hour. There, the Milky Way's larger spiral companion – Messier 31, the Andromeda galaxy – awaits, sitting directly in the Milky Way's path of travel.

The effects of the collision will begin millions of years before the two galaxies actually touch. In the same way that the Moon's tidal influence gently pulls Earth's oceans away from the sea floor, the Milky Way's leading edge will begin to fall faster towards Andromeda than its trailing side. A tail thousands of light years long will stretch outwards as the galaxies strike, creating a myriad of tidal loops, rings and streams. Eventually, they will coalesce into a single, disrupted ellipsoidal galaxy system. The entire collision process will last more than a billion years.

Despite the breakneck speed at which they hit, the Milky Way and Andromeda are so large that it will be 100 million years after their spiral arms first make contact before their galactic centres pass one another. Meanwhile, the spaces between the stars in each galaxy are so vast – imagine a tennis ball at Wimbledon zipping past another tennis ball at Roland Garros – they will all fly past one another like two huge swarms of honeybees. The stars don't need to touch, though, for their solar systems to be devastated; the wildly varying gravitational pulls in every direction could disrupt the orbit of almost every planet and moon, likely either driving them into their suns or flinging them out into interstellar space.

It is currently impossible to know the fate of the Sun and its planets when the Milky Way–Andromeda collision happens. The view into the sky, though, will certainly be spectacular, and not only because of all the stars streaming around. Clouds of cold interstellar gas in the galaxies will fall together, sending shock waves through each other. The glowing shock fronts stretching across the sky will crash through the clouds, triggering their collapse and sparking the birth of new stars. The interstellar light show should last for millions of years.

In 2017, a kilometre-long object flew through our solar system, tumbling in at high speed during the summer and exiting in late autumn. Named 'Oumuamua (Hawai'ian for 'scout'), it could have been an asteroid thrown out of its original solar system millions of years ago by a close stellar encounter.

BIOGRAPHY

In 1924, the American astronomer **Edwin Hubble** (1889–1953) used Henrietta Swan Leavitt's period-luminosity law to measure the distance to the Andromeda galaxy. Using the 2.5-metre (100-inch) Hooker telescope on Mount Wilson, California, he measured the properties of Cepheid variable stars he found in Andromeda to show that it was more than 1 million light years away from Earth and thus, like the Milky Way, a vast galaxy filled with billions of stars.

5,000,000,000 *years from now*

The Andromeda galaxy (also known as Messier 31) is currently about 2,200,000 light years away from Earth. Like the Milky Way, Andromeda has a number of smaller galaxies orbiting it, including the dwarf elliptical galaxy Messier 32 and the Triangulum spiral galaxy (Messier 33). The Milky Way, Andromeda and their satellite galaxies together comprise what astronomers call the 'Local Group' of galaxies — which, even with the Milky Way and Andromeda falling towards one another, is as a whole falling in the direction of the constellation Virgo at more than half a million kilometres per hour (300,000 miles per hour). The image below shows how our night sky may appear 3,750,000,000 years from now, as Andromeda approaches and causes tidal effects that distort the shape of the Milky Way.

7,000,000,000 *years from now*

Atoms Disintegrate and Black Holes Evaporate

As described by the general theory of relativity, space and time are linked in a vast, flexible four-dimensional fabric.

With that structure as its foundation, perhaps the most fundamental way to mark the passage of cosmic time has been to observe the expansion of the universe, which has been going on since the Big Bang. But how far into the future can we use this temporal yardstick? In other words, will the universe continue to expand forever?

A quarter of a century ago, astronomers finally began to obtain the highly precise observational data we needed to answer that ultimate question. A careful measurement of how fast our corner of the universe is expanding, combined with increasingly reliable calculations of the amount of matter in the universe, confirmed that cosmic expansion would indeed continue indefinitely. So the question shifted: what will happen to the contents of the universe as time rolls onwards towards eternity?

Familiar events currently taking place will be the first to finish. The birth of galaxies, stars, planets and life in the universe all depend on having the raw materials available to build them. When every available atom of gas and every grain of dust that can serve as fuel or fodder is exhausted, the formation of new objects will cease. Existing objects will live out their lives, and the universe will gently go dark. The best scientific estimates show that the majority of cosmic gas is still available, and the universe will go on for at least 10,000 times longer than its current age before the stars go out. That's 1,000 trillion years from now.

The last remaining concentrations of matter after the end of stars will be the stellar remnants: brown dwarfs, white dwarfs, neutron stars and black holes will populate the universe where stars once shone. How long will they stay around? According to the leading theories of particle physics, the protons that comprise them will one day decay, and the remnants themselves will disintegrate. Proton decay hasn't been experimentally confirmed yet, though. Even if it does occur, it won't even be noticeable until at least a billion trillion trillion years from now.

And what about black holes – the ultimate manifestations of cosmic structure? Even these most permanent of objects will likely disappear. Theoretical calculations have shown that over a nearly immeasurably long time, tiny amounts of radiation can seep out of black holes. A black hole the mass of the Sun would take about a trillion trillion trillion trillion trillion trillion years to evaporate completely. Supermassive black holes, like the one at the centre of the galaxy Messier 87, which weighs in at more than 6 billion Suns, would take at least a trillion trillion times longer than that. On a piece of paper, write down a one, and then write a hundred zeroes after it; that's about the number of years it will take for all the black holes in the universe to evaporate.

One of the original key project science goals of the Hubble Space Telescope was achieved in 2000, when an international team of astronomers led by Canadian Wendy Freedman (b. 1957), American Robert Kennicutt (b. 1951) and Australian Jeremy Mould (b. 1949) made the most precise measurements ever of Cepheid variable stars outside our Milky Way galaxy. Using the Leavitt Law that describes the period-luminosity relation of Cepheids, they measured the rate of cosmic expansion and showed it will likely go on forever into the future.

For most of the twenty-first century so far, physicists have been using the Super-Kamiokande neutrino observatory in Gifu Prefecture, Japan, to search for signs of proton decay. In the coming decade, an upgraded facility called Hyper-Kamiokande will begin operation that will be about 10 times more sensitive.

BIOGRAPHY

English astrophysicist **Stephen Hawking** (1942–2018) applied quantum theory to the study of black holes. He showed that energy could ooze very slowly out of a black hole, gradually reducing its mass and size, leading to its eventual disappearance.

Dark Energy and (maybe) the End of Time

Every object in the universe is gone. The cosmos continues on in final, perfect stillness. Or not.

Until the end of the twentieth century, when astronomical observations showed that the universe would expand forever, theoretical models had predicted the possibility that, if there were enough dark matter in the universe, cosmic expansion might eventually slow down, stop and reverse. The result would be a 'big crunch' – a sort of anti-Big Bang – and the universe would disappear into a black hole-like singularity. In that scenario, during the contraction, the usual method of measuring the passage of cosmic time would be running the wrong way; that is, we would all be travelling backwards in time.

Although that 'closed universe' scenario has been ruled out by scientific data, our actual end result may prove to be even more dramatic. For billions of years, the speed at which the universe is expanding has been slowly but measurably going up. Furthermore, like an inflating pool float, the universe should be changing shape as it grows – but it isn't; across the cosmos, the dimensions of length, width and height haven't bent at all. To use the astronomical terminology, the geometry of space-time is flat.

Amazingly, these measurements imply that more than two-thirds of all the energy in the universe has been driving cosmic expansion, and making that expansion faster and faster. Unlike familiar forms of energy such as heat and light, that prodigious 'dark energy' seems not to interact with matter; yet its presence in space is three times denser than even the cosmological 'dark matter' that itself is five times more plentiful than ordinary baryonic matter. If we add it all up, only 5 per cent of the matter and energy in the universe is comprised of things we currently scientifically understand.

What does it all mean? Humanity's scientific quest has brought us quite a way so far. For a single species of carbon-based life living on a rocky planet, orbiting an ordinary star in the disc of an unremarkable galaxy, we have learned a great deal about the history of our universe, where we are in it and what will happen to it far into the future. Even so, after all the black holes and stars and planets and protons are gone, even if the universe seems silent and dark, 95 per cent of the contents of the universe may well still be there – and those contents remain completely mysterious to us. Do dark matter and dark energy harbour the potential to brighten that quiet night? Or, just as physical symmetries broke in the moments after the Big Bang, triggering the creation of all that exists today, will conditions then be right for a new, even more amazing cosmic age?

We have arrived at the end of our journey through cosmic time. The conclusion of every journey heralds the beginning of a new adventure. Our scientific curiosity, creativity and ingenuity will lead us forwards. Awesome discoveries await!

We all travel forwards through the dimension of time. Can anything travel backwards in time? Although this idea has caught the fancy of both storytellers and scientists throughout history, travel to the past appears to violate important symmetries and conservation laws that govern the behaviour of our physical universe. Then again, symmetries have been broken before; and a few conservation laws once thought immutable have been experimentally shown to be wrong, leading to new discoveries about the way our cosmos works.

BIOGRAPHY

Chinese-American physicist **Chien-Shiung Wu** (1912–1997) was the only scientist of Chinese descent known to have worked on the Manhattan Project to develop the first atomic bomb. In 1956, she and her graduate student Marion Biavati showed experimentally that parity (the property that particles follow the same rules as their mirror images) is not always conserved in subatomic reactions. Parity violation is a fundamental asymmetry that may hold clues about the possibility of travelling backwards in time, and it highlights the many scientific mysteries we have yet to unlock in our understanding of space and time.

Glossary

accretion disc A flattened disc of gaseous material centred on a strong gravitational source such as a planet, star or black hole.

antimatter Massive particles (such as positrons and antiprotons) analogous to typical matter but with opposite electrical charge.

asteroid A rocky or metallic object in space between about 0.1 and 1,000 kilometres (0.06 and 600 miles) in size.

atom The basic component of a chemical element, composed of a nucleus and one or more electrons.

baryon A subatomic particle composed of gluons and three (or sometimes more) quarks.

baryonic matter Matter that is dominated by baryons such as protons, neutrons, atoms and molecules; stars, planets and people are made of baryonic matter.

binary star/binary system A star system that contains two stars orbiting one another.

black hole A region of space-time where gravity is so extreme that its escape velocity exceeds the speed of light.

boson A type of subatomic particle that carries force between other particles.

cosmic microwave background Electromagnetic radiation left over from the early universe that fills all of space.

dark energy An as-yet-unidentified form of energy that appears to be causing the expansion of the universe to accelerate.

dark matter Cosmic matter that is not composed of baryons, emits no electromagnetic radiation and interacts with baryonic matter only via gravity.

deuteron An atomic nucleus consisting of one proton and one neutron.

DNA Short for deoxyribonucleic acid, a kind of molecule that Earth's carbon-based life forms use to reproduce and preserve genetic information.

electron A lepton that has negative electrical charge and is a major component of atoms.

eukaryote Life forms composed of cells with DNA inside a nucleus; humans are eukaryotes.

event horizon The boundary around a black hole within which nothing, including light, can escape.

evolution The process of species or categories of organisms or systems changing over time; life on Earth has experienced evolution by natural selection.

exoplanet A planet outside our solar system, usually orbiting stars other than the Sun.

fermion A type of subatomic particle that has mass and volume.

fundamental forces The four basic kinds of force in the universe: gravity, the weak nuclear force, electromagnetism and the strong nuclear force.

general theory of relativity The scientific theory that mathematically describes how space, time and gravity are related in the cosmos.

gluon A boson (there are eight kinds) that carries the strong nuclear force.

gravitational waves Ripples of space caused by extremely energetic cosmic events such as the collision of two black holes.

Hawking radiation A physical process, first proposed by Stephen Hawking, whereby black holes could lose their mass and evaporate after a very long period of time.

Higgs boson A subatomic particle that facilitates interactions between other particles and the Higgs field.

Higgs field A potential field that fills the universe and interacts with particles to produce mass.

hominid A type of life on Earth that includes modern human beings and their extinct genetic relatives.

ionization A process that causes matter to become electrically charged.

isotope A version of a chemical element with a specific number of neutrons and protons in its atomic nuclei.

lepton A type of fermion (there are six kinds) that includes electrons and neutrinos.

light year The distance light travels through the vacuum of space in one year, about 9.5 trillion kilometres (5.9 trillion miles).

mass The property of matter that makes it heavy and thus move slower than the speed of light.

meteorite A solid object, usually icy or rocky or metallic, that strikes Earth's surface from outer space.

nebula A cloud of gas and interstellar dust in outer space that often either contains very young stars or is produced by old or dying stars.

nucleosynthesis (or nuclear fusion) A physical process that combines lighter, simpler atomic nuclei to form heavier, more complex atomic nuclei.

neutrino A type of lepton that has very little mass – not to be confused with a neutron, which is a type of baryon.

neutron star A type of extremely dense stellar remnant, usually the result of a supernova explosion.

photon A type of boson that carries electromagnetic force; light is composed of photons.

Planck time The earliest period of cosmic history, so short that within it the laws of physics are not yet understood.

potential field An invisible region of space, usually surrounding an object, that can cause energy or force to be produced when certain kinds of particles interact with it.

proton The most common baryon in the universe; it is the nucleus of a hydrogen atom, and it contains gluons, two up quarks and one down quark.

pulsar A neutron star that spins very fast and produces regular pulses of radiation as it spins.

quantum fluctuation A sudden and very short-lived change in the amount of energy contained in one tiny part of the universe.

quantum mechanics The scientific theory that mathematically describes the physical behaviour of atomic and subatomic particles.

quark A type of fermion (there are six kinds) that are the building blocks of baryons and other complex particles.

quasar A very bright source of energy and radiation powered by a supermassive black hole at its centre.

radioactive decay A process where heavier, more complex atomic nuclei shed subatomic particles and become lighter, less complex nuclei over time.

recombination A process that is the opposite of ionization, as charged particles come together to form neutral atoms.

red giant A stage of the life cycle of stars; the Sun will become a red giant in about 5 billion years' time.

singularity A single location in space and time where the standard rules of physics do not apply, such as in a black hole or at the Big Bang.

space-time The four-dimensional 'fabric' in which we and all the other constituents of the universe exist.

special theory of relativity The scientific theory that mathematically relates how objects move through space and time at different speeds.

subatomic particle A particle, such as a boson or fermion or baryon, which is smaller than an atom.

supernova A powerful explosion produced at the end of a star's life cycle.

uniformitarianism The scientific idea that the laws of nature operate throughout the universe the same way in the past, present and future, leading to slow and gradual changes to complex systems such as the planet Earth.

white dwarf A type of hot stellar remnant; the Sun will be a white dwarf in about 7 billion years' time.

Resources

Books

30-Second Universe
Charles Liu, Karen Masters, Sevil Salur
(IVY PRESS, 2019)

Brief Answers to the Big Questions
Stephen Hawking
(BANTAM, 2018)

Cosmos
Carl Sagan
(RANDOM HOUSE, 1980)

The Diversity of Life
Edward O. Wilson
(W. W. NORTON, 1999)

Einstein's Monsters: The Life and Times of Black Holes
Chris Impey
(W. W. NORTON, 2018)

The End of Everything (Astrophysically Speaking)
Katie Mack
(SCRIBNER, 2021)

The First Human: The Race to Discover Our Earliest Ancestors
Ann Gibbons
(DOUBLEDAY, 2006)

The First Three Minutes: A Modern View of the Origin of the Universe
Steven Weinberg
(BASIC BOOKS; 2ND EDN,1993)

The Handy Astronomy Answer Book
Charles Liu
(VISIBLE INK PRESS; 3RD EDN, 2013)

The Handy Physics Answer Book
Charles Liu
(VISIBLE INK PRESS; 3RD EDN, 2020)

Reality in the Shadows (or) What the Heck's the Higgs?
S. James Gates Jr, Frank Blitzer and Stephen Jacob Sekula
(YBK, 2017)

Until the End of Time: Mind, Matter, and Our Search for Meaning in an Evolving Universe
Brian Greene
(KNOPF, 2020)

Warped Passages: Unravelling the Mysteries of the Universe's Hidden Dimensions
Lisa Randall
(ECCO, 2005)

The Zoologist's Guide to the Galaxy
Arik Kershenbaum
(VIKING-PENGUIN, 2020)

Websites

AAS – American Astronomical Society
aas.org
The major organization of professional astronomers
and astrophysicists in North America, founded in 1899,
headquartered in Washington, DC.

AGU – American Geophysical Union
agu.org
Dedicated since 1919 to the advancement of Earth
and space science, with more than 60,000 members
in 137 countries.

CERN – European Organization for Nuclear Research
home.cern
The largest particle physics laboratory in the world,
located at the French-Swiss border and home to the
Large Hadron Collider (LHC).

ESA – European Space Agency
esa.int
Intergovernmental organization of 22 member nations
dedicated to space science and exploration, with
headquarters in Paris and a spaceport in French Guiana.

Geological Society of London
www.geolsoc.org.uk
The United Kingdom's national society for geoscience,
founded in 1807, with the aim of improving knowledge
and understanding of the Earth.

IAU – International Astronomical Union
iau.org
Association of professionals active in research
and education in astronomy, founded in 1919
with its headquarters in Paris.

The Leakey Foundation
leakeyfoundation.org
Formed in 1968 to increase scientific knowledge,
education and public understanding of human origins,
evolution, behaviour and survival.

RAS
Royal Astronomical Society
ras.ac.uk
Learned society and charity headquartered in London,
founded as the Astronomical Society of London in 1820.

STScI – Space Telescope Science Institute
stsci.edu
Home institution, based in Baltimore, of the Hubble
Space Telescope, James Webb Space Telescope, and the
future Nancy Grace Roman Telescope.

Universe 101
map.gsfc.nasa.gov/universe/
An online primer created by NASA scientists and
educators on cosmology and the Big Bang.

Vera C. Rubin Observatory
lsst.org
An international project with its main telescope in Chile
that over the next decade will produce the deepest
and widest image of the universe in human history.

Index

ABOUT THE AUTHOR

Charles Liu is a professor and chair of the Department of Physics and Astronomy at the City University of New York's College of Staten Island, and an associate astrophysicist with the American Museum of Natural History and Hayden Planetarium in New York. His research focuses on colliding galaxies, supermassive black holes and the star formation history of the universe. He hosts the podcast *The LIUniverse with Dr. Charles Liu* and has authored nine books about the physical and astronomical sciences. He earned degrees at Harvard and the University of Arizona, and held postdoctoral appointments at Columbia University and Kitt Peak National Observatory. He currently serves as president of the Astronomical Society of New York, and in 2020 he was named a Legacy Fellow of the American Astronomical Society. Charles and his wife, Dr. Amy Rabb-Liu, have three cosmically curious children.

Acknowledgements

The author extends heartfelt thanks to the team of editors, especially Caroline Earle, whose work made this book possible. Support for the author during the course of this work was provided by the US National Science Foundation, the City University of New York, the Alfred P. Sloan Foundation and NASA.

Picture credits

Key: t=top; b=bottom; l=left; r=right, c=centre; and variations thereof.

Cover illustration by Ted Jennings
Inside illustrations by Maksim Malowichko.
All other images:

2 & 103 NASA,ESA, M. Robberto (Space Telescope Science Institute/ESA) and the Hubble Space Telescope Orion Treasury Project Team; 4–5 & 135 NASA/CXC/M. Weiss; 52–53 NASA/WMAP Science Team; 67 t NASA, ESA, and P. Oesch (Yale University); 67 b ALMA (ESO/NAOJ/NRAO), NASA/ESA Hubble Space Telescope, W. Zheng (JHU), M. Postman (STScI), the CLASH Team, Hashimoto et al; 70 & 81 NASA/JPL-Caltech/ESO/R. Hurt; 74 NASA, ESA, D. Elmegreen (Vassar College), B. Elmegreen (IBM's Thomas J. Watson Research Center), J. Sánchez Almeida, C. Munoz-Tunon & M. Filho (Instituto de Astrofísica de Canarias), J. Mendez-Abreu (University of St Andrews), J. Gallagher (University of Wisconsin-Madison), M. Rafelski (NASA Goddard Space Flight Center) & D. Ceverino (Center for Astronomy at Heidelberg University); 75 NASA, ESA, H. Teplitz and M. Rafelski (IPAC/Caltech), A. Koekemoer (STScI), R. Windhorst (Arizona State University), and Z. Levay (STScI); 77 NASA, ESA, S. Baum and C. O'Dea (RIT), R. Perley and W. Cotton (NRAO/AUI/NSF), and the Hubble Heritage Team (STScI/AURA); 79 tl NASA, Holland Ford (JHU), the ACS Science Team and ESA; 79 tr NASA, ESA, and The Hubble Heritage Team (STScI/AURA); 79 cl NASA, ESA, S. Beckwith (STScI), and The Hubble Heritage Team (STScI/AURA); 79 cr ESA/Hubble & NASA, A. Bellini; 79 bl NASA, ESA, and The Hubble Heritage Team (STScI/AURA); 79 br ESA/Hubble & NASA, J. Lee and the PHANGS-HST Team. Acknowledgement: Judy Schmidt; 87 NASA, ESA, and the Hubble Heritage (STScI/AURA)-ESA/Hubble Collaboration; 89 t T.A. Rector (University of Alaska Anchorage), Richard Cool (University of Arizona) and WIYN; 89 b NASA, ESA, and the Hubble Heritage (STScI/AURA)-ESA/Hubble Collaboration;

Acknowledgment: J. Mack (STScI) and G. Piotto (University of Padova, Italy); 105 t ESA/Hubble and NASA; acknowledgement: Judy Schmidt; 105 b ALMA (ESO/NAOJ/NRAO), NSF; 107 Mercury NASA/Johns Hopkins University Applied Physics Laboratory/Carnegie Institution of Washington; 107 Venus NASA/JPL; 107 Earth NASA; 107 Mars NASA, ESA, the Hubble Heritage Team (STScI/AURA), J. Bell (ASU), and M. Wolff (Space Science Institute); 107 Jupiter NASA/ESA/NOIRLab/NSF/AURA/M.H. Wong and I. de Pater (UC Berkeley) et al.; Acknowledgments: M. Zamani; 107 Saturn NASA/JPL-Caltech/Space Science Institute; 107 Uranus Lawrence Sromovsky, University of Wisconsin-Madison/W.W. Keck Observatory; 107 Neptune NASA/JPL; 109 t NASA/ESA/H. Weaver and E. Smith (STScI); 109 b H. Hammel, MIT and NASA; 119 USGS/Public domain; 123 Frank Fox/http://www.mikro-foto.de (CC BY-SA 3.0 DE); 125 t Dr. Norbert Lange/Shutterstock; 125 cl Gertjan Hooijer/Shutterstock; 125 cr Henri Koskinen/Shutterstock; 125 b Eugen Haag/Shutterstock; 126 & 139 Walter Myers/Stocktrek Images/Getty Images; 140 & 155 NASA, ESA, and R. Kirshner (Harvard-Smithsonian Center for Astrophysics and Gordon and Betty Moore Foundation) and P. Challis (Harvard-Smithsonian Center for Astrophysics); 143 NASA, ESA, J. Hester, A. Loll (ASU); 147 l ESA/NASA; 147 r SOHO-EIT Consortium, ESA, NASA; 153 NSF/LIGO/Sonoma State University/A. Simonnet; 156 NASA; 159 NASA; 160 l Frank Nowikowski/Alamy Stock Photo; 160 r–161 Ikkel Juul Jensen/Science Photo Library; 163 t Pics by Nick/Shutterstock; 163 cl Guillaume Blanchard (CC BY-SA 3.0); 163 cr Marie-Lan Nguyen (CC BY 2.5); 163 b thipjang/Shutterstock; 165 tl Realy Easy Star/Alamy Stock Photo; 165 tc El Greco 1973/Shutterstock; 165 tr Daniel Schwen (CC BY-SA 4.0); 165 c Andrew Roland/Shutterstock; 165 b travelview/Shutterstock; 167 (1 to 6 clockwise from tl): 1: Rurik the Varangian (public domain), 2: Courtesy of Science History Institute, 3: NASA, 4: NASA/JPL-Caltech/ASU, 5: Wright Stuf (public domain), 6: Ttaylor (public domain); 169 t NASA/JPL-Caltech/SETI Institute; 169 b NASA/JPL-Caltech/MSSS; 175 NASA Earth Observatory images by Joshua Stevens, using Suomi NPP VIIRS data from Miguel Román, NASA's Goddard Space Flight Center; 177 SOHO-EIT Consortium, ESA, NASA; 179 NASA, ESA, Z. Levay and R. van der Marel (STScI), T. Hallas, and A. Mellinger.